FACIAL GEOMETRY

ABOUT THE AUTHOR

Dr. Robert M. George is a member of the biology faculty at Florida International University where he teaches human and comparative anatomy, primate and human evolution, and forensic osteology. He received his doctorate in physical anthropology from the University of Washington and has held appointments in several medical schools including Brown University, the University of Kuwait, St. George's University (Grenada), The University of Chicago and the University of South Carolina. In 1985, he began research in the field of forensic facial approximation and has served as a consultant on numerous forensic cases from coast-to-coast including the notorious "Green River Killer" case. His professional memberships have included the American Association of Anatomists, the American Association of Physical Anthropologists, the American Academy of Forensic Sciences and the Florida Division of the International Association for Identification where he currently serves as the chairperson of the Forensic Composite Art Committee. When not teaching, consulting, or approximating, he enjoys portraiture and has a passion for caricaturing.

FACIAL GEOMETRY

Graphic Facial Analysis for Forensic Artists

By

ROBERT M. GEORGE, Ph.D.

Department of Biology
Florida International University

CHARLES C THOMAS • PUBLISHER, LTD.
Springfield • Illinois • U.S.A.

Published and Distributed Throughout the World by

CHARLES C THOMAS • PUBLISHER, LTD.
2600 South First Street
Springfield, Illinois 62704

©2007 by CHARLES C THOMAS • PUBLISHER, LTD.

ISBN 978-0-398-07770-9 (spiral)

Library of Congress Catalog Card Number: 2007016975

*With THOMAS BOOKS careful attention is given to all details of manufacturing
and design. It is the Publisher's desire to present books that are satisfactory as to their
physical qualities and artistic possibilities and appropriate for their particular use.
THOMAS BOOKS will be true to those laws of quality that assure a good name
and good will.*

Printed in the United States of America
SR-R-3

Library of Congress Cataloging-in-Publication Data

George, Robert M.
   Facial geometry : graphic facial analysis for forensic artists / Robert M. George.
      p. cm.
   Includes bibliographical references.
   ISBN 978-0-398-07770-9 (pbk.)
   1. Facial reconstruction (Anthropology) 2. Face--Measurement. 3. Face--Imaging.
4. Face--Identification. I. Title.

GN74.G46 2007
614'.17--dc22

                                  2007016975

*To my loving wife, Mary, who is truly golden in the
eyes of this beholder.*

BEAUTIFUL QUOTES

The human features and countenance, although composed of but some ten parts or little more, are so fashioned that among so many thousands of men there are no two in existence who cannot be distinguished from one another.

Pliny the Elder (circa 50 A.D.)

It is the common wonder of all men, how among so many millions of faces there should be none alike.

Thomas Browne

Beauty is the opposite of deformity.

Leonardo

There is no excellent beauty which hath not some strangeness of proportion.

Francis Bacon

Beauty is (altogether) in the eye of the beholder.

Margaret Hungerford
(Lew Wallace)

No standard rules are included for those average heads which don't exist.

Louise Gordon

If the beauty of Helen of Troy was powerful enough to sink a thousand ships, then that force capable of sinking a single ship would be a milliHelen.

Anonymous

FOREWORD

Robert George, B.A., M.S., PhD., anatomist,
physical anthropologist and artist.

That is my memory of his personal card about twenty years ago given to me after his first lecture at the Facial Reconstruction Seminar held then on Wednesday evening at the annual meeting of the American Academy of Forensic Sciences.

After twenty years of professional and personal contact with Dr. George and watching his career productivity, the inescapable conclusion is that card announced the special power of those disciplines all rolled into one great mind.

In being a clinical orthodontist, similarly, my interest is the face, particularly the growing face and what can be done to modify facial bones along with tooth movement so that my patient's final records, both radiographic, cephalometric and photographic, illustrate how close I can strong-arm the anatomy to fit our anatomic and esthetic measured ideals for facial and dental proportion.

However, in forensic dentistry and anthropology, the quest is not for measuring sameness, if you will, but rather for dissecting out the characteristics that make the person different from all others, as in the human identification protocol. My collaboration with Dr. George involved utilization of these records for research.

Dr. George has used his own built-in basic multidisciplinary talent to add real science to the study of the relationship of hard tissue to soft tissue in the face, particularly radiographic. Knowing limitations of technique and the statements that can and cannot be made from scientific studies, his terminology "facial approximation" is a learned professional contribution.

Dr. Arthur Burns
Jacksonville, FL
2007

PREFACE

Forensic art may be defined as "portrait art minus a tangible subject." Whether rendering a composite illustration as dictated by a witness or attempting a facial approximation (reconstruction, reproduction, etc.) on an unidentified skull, the primary goal of the forensic artist remains the same—to depict as accurately as possible the overall shape and contouring of the face including its ciliary adornments. While the precise details of eye, nose and lip anatomy are extremely difficult to ascertain from verbal descriptions or bony substrates, it is possible to accurately position these features and to relate them via numerous craniofacial indices.

Cephalometry, the measurement of the head and face, is a centuries-old science developed by anatomists, anthropologists, artists, maxillofacial surgeons, orthodontists and others interested in facial esthetics. The subdiscipline of graphic facial analysis (GFA) is the quantitative assessment of the relationships of facial features as determined by specific indices and angles. For example, a nose may be judged to be long with reference to total facial length, the mouth to be wide relative to facial width, the mandible to be prognathic as determined by the mandibulofacial angle, and so on. An understanding of GFA is essential for composite drawings, forensic facial approximations and photographic comparisons.

Forensic art came of age in 2001 with the publication of Karen Taylor's masterwork, *Forensic Art and Illustration*. This book covered all aspects of forensic art in 18 chapters and a weighty 561 pages. Another excellent reference work is *Forensic Facial Reconstruction* by Caroline Wilkinson (2004) and newer computer techniques have been introduced by John Clement and Murray Marks in their book *Computer-Graphic Facial Reconstruction* (2005). The main objective of the present more modest manual is to present a series of practical indices interrelating the key features of the human face that will provide a foundation for any exercise in forensic art from composite sketch to post-mortem "refacing." These indices will be illustrated with a survey of the numerous and often surprising geometric forms that permeate facial design. We will examine the various triangles and rectangles, rhomboids and trapezoids, parallelograms and circles that on the one hand define the human face (the theme) and on the other, give it its individuality (variations on the theme).

This book deals purely with the measurement of facial variation. It is **not** a discourse on esthetics or the analysis of beauty which has been a major entertainment over the centuries. Ever since Phidias (5th century B.C.) created "golden geometry" with his golden ratio (a.k.a. the divine proportion), artists and mathematicians have been scrutinizing the human face for its "golden plan." The underlying assumption is that the more golden relations, rectangles or triangles a face possesses the more beautiful it will be. This theme was developed by the mathematician Matila Ghyka in her fascinating book *The Geometry of Art and Life* (1946). Ricketts (1982) and Shoemaker (1987) used similar analyses in their studies of dental esthetics and the maxillo-facial surgeon, Dr. Stephen Marquardt (2002), has constructed the ultimate golden mask purportedly underlying the essence of facial beauty. In this book, I am not concerned with value judgments since I have no way of knowing if 1.618 is actually prettier than 3.141 or 6.022.

Neither is this book intended as a clinical reference. Clinical cephalometry was completely covered by Farkas and Munro in their classic *Anthropometric Facial Proportions in Medicine* (1987) with more recent applications provided by Kolar and Salter in *Craniofacial Anthropometry* (1997). These are works of solid research and sound scholarship and provide the statistical data base followed in the present book.

This book is divided into four chapters. Chapter 1 defines the cephalometric points, planes, areas and lines that demarcate the human face. The detailed surface anatomy of the eye, nose, mouth and ear is also included. Chapter 2 reveals the underlying geometry of the human facial plan. A selection of triangles, rectangles and other polygons are illustrated, some "golden" and some less lustrous. Chapter 3 covers the graphic facial analysis (GFA) of the frontal face. Sixteen indices and triangles are defined and illustrated with their means and ranges of variation. Chapter 4 details the GFA of the lateral face by means of 8 angles and indices with special attention given to the nose and ear.

Forensic art has long suffered from a lack of standardization. Not only do forensic artists vary in skill from rank amateurs to professional portraitists, they also use different methods, proportion templates and tissue thickness tables. The major objective of this book is to provide a working set of facial indices that have been statistically evaluated for male and female Caucasians as a step toward such standardization. Equivalent data for people of African and Asian ancestry is singularly lacking and is an open field for future research.

R.M.G.

CONTENTS

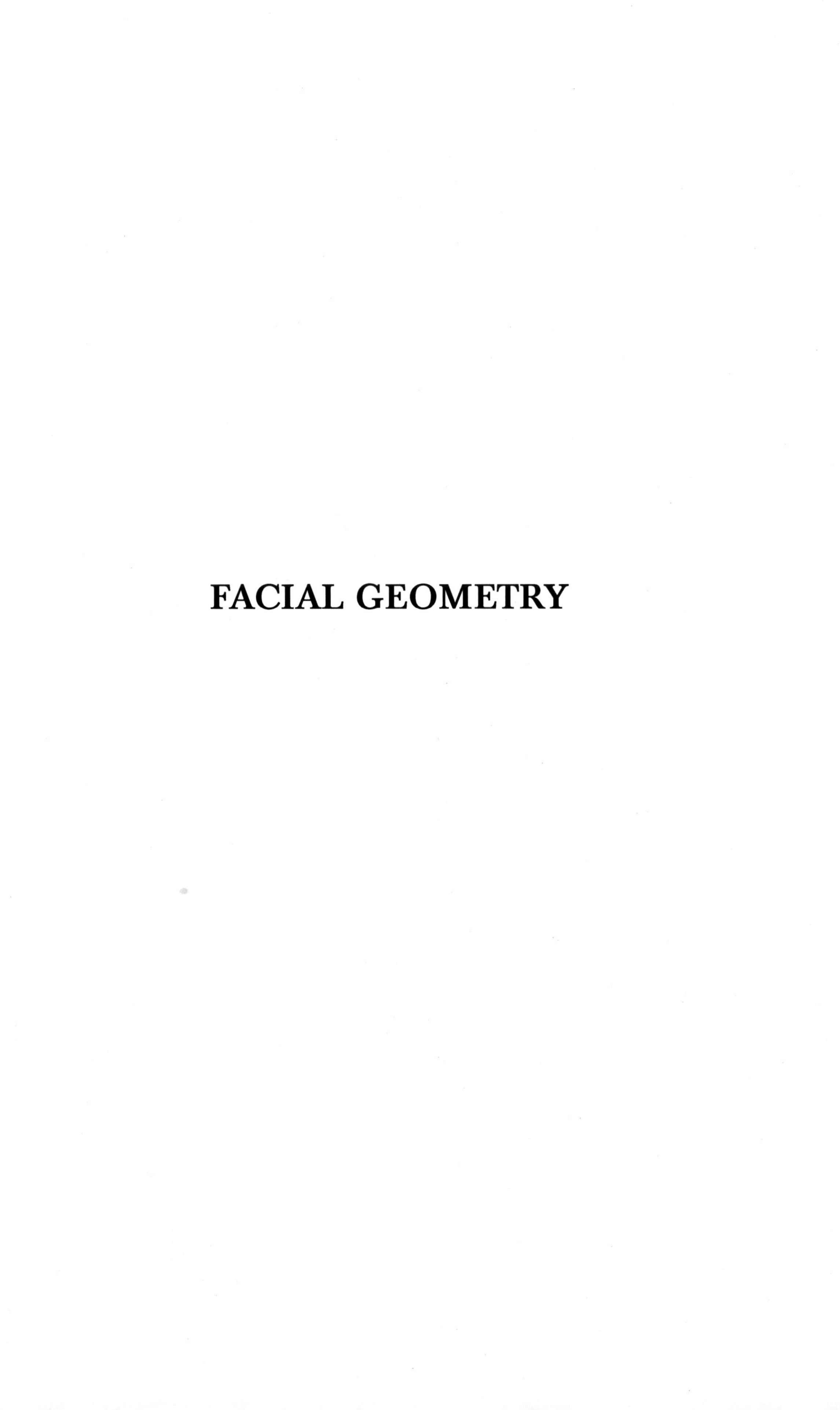

FACIAL GEOMETRY

Chapter 1

FACIAL GEOGRAPHY

The human face is a marvel of contouring displaying hills and valleys, slopes, crests, ridges and crevasses all wrapped around a spherical skull. Throughout this irregular terrain there are certain constant landmarks which are denoted by an array of precisely defined cephalometric points. These points allow us to map the geography of the face, to zone its areas and ultimately, to develop indices by which subtle relationships may be revealed. Facial cephalometric points correspond to underlying skeletal craniometric points and a knowledge of their correlations forms the scientific basis of forensic facial approximation (George, 1987, 1993). The most significant of these facial points are defined in the following section and illustrated in Figures 1.1 and 1.2.

CEPHALOMETRIC POINTS

Cranial Points

- **vertex (v)**–the midline apex of the neurocranium (braincase).
- **supraglabella (sg)**–an arbitrary point about one inch above the glabella (used in forensic facial approximations).
- **glabella (g)**–the "bald spot" between the eyebrows on the midsagittal plane (MSP).
- **euryon (eu)**–the most lateral point of the neurocranium in the parietotemporal area, best plotted from the frontal view.
- **auriculotemporale (at)**–the superior junction of the ear with the side of the head; this point is highly variable, subject to rotational distortions and usually obscured by hair; it is significant when plotting certain facial triangles (new term proposal).

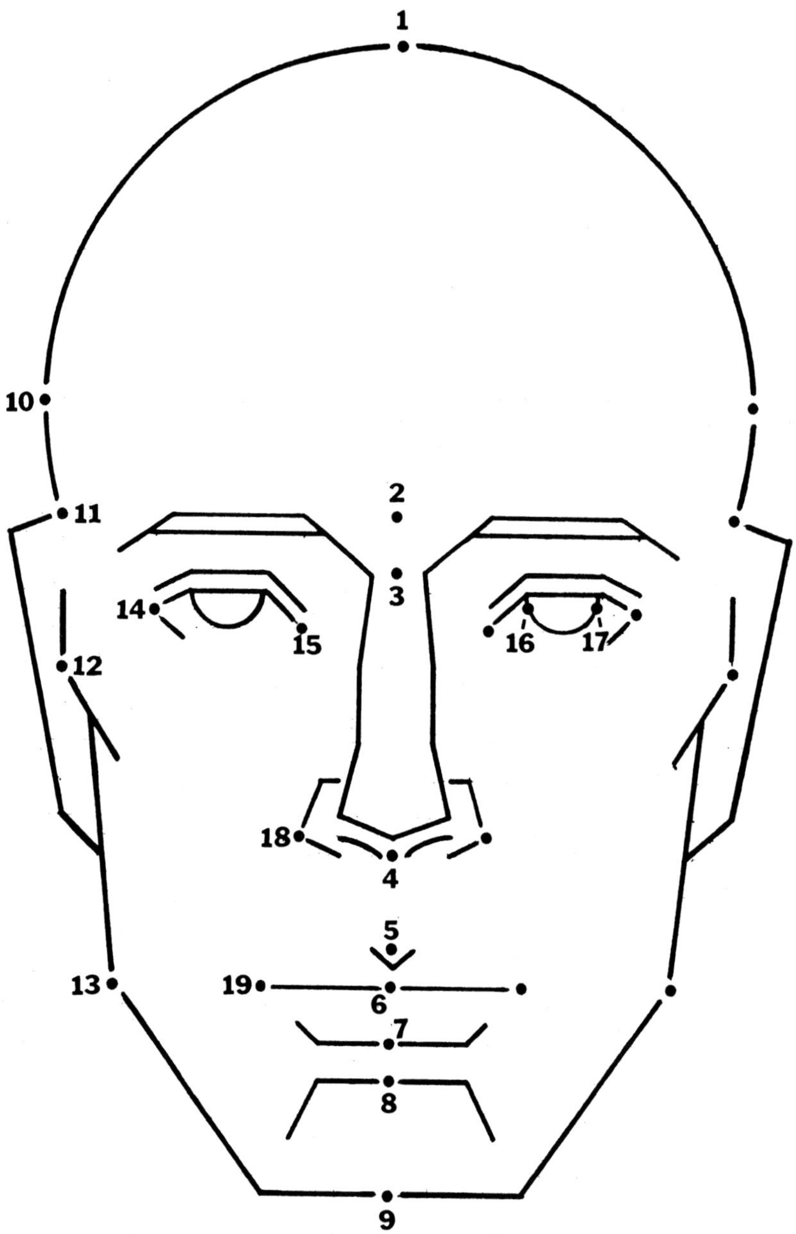

Figure 1.1. Cephalometric Points - Frontal View: 1 = vertex; 2 = glabella; 3 = nasion; 4 = subnasale; 5 = labiale superius; 6 = stomion; 7 = labiale inferius; 8 = labiomentale; 9 = gnathion; 10 = euryon; 11 = auriculotemporale; 12 = zygion; 13 = gonion; 14 = ectocanthion; 15 = endocanthion; 16 = iridion mediale; 17 = iridion laterale; 18 = alare; 19 = chelion.

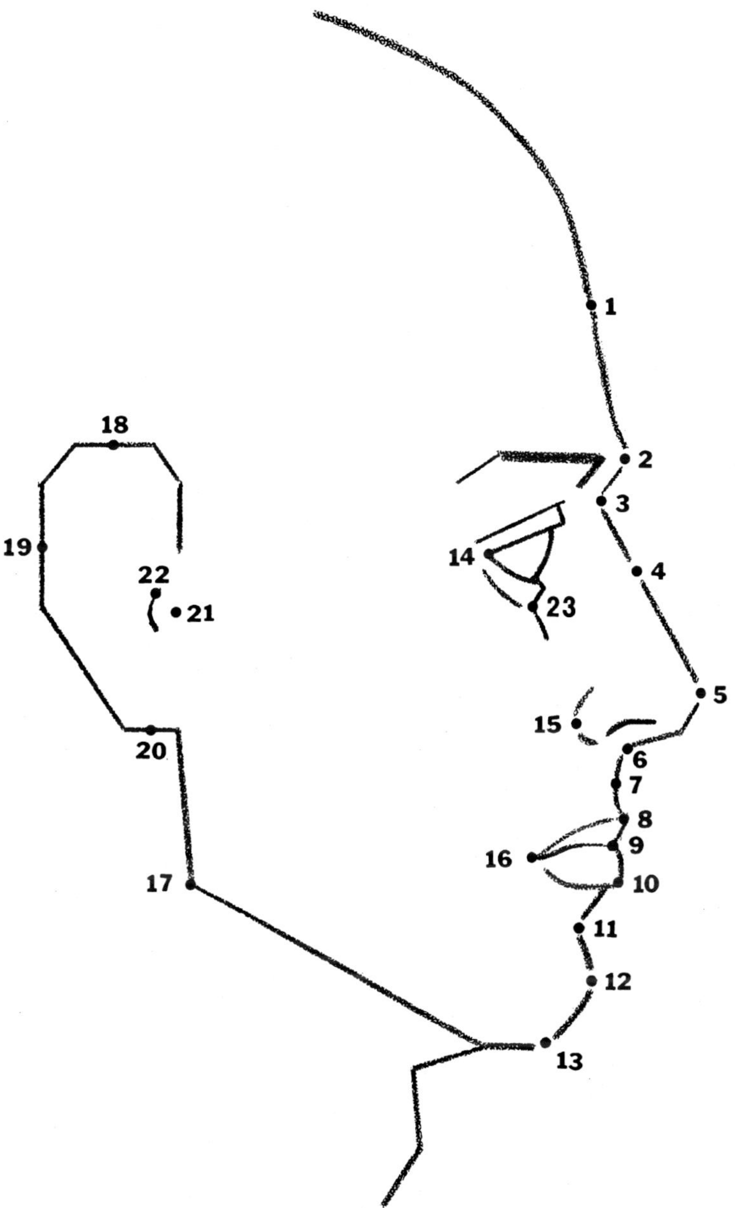

Figure 1.2. Cephalometric Points - Lateral View: 1 = supraglabella; 2 = ciliare (glabella); 3 = nasion; 4 = nasale; 5 = pronasale; 6 = subnasale; 7 = superior labial sulcus; 8 = labiale superius; 9 = stomion; 10 = labiale inferius; 11 = labiomentale (inferior labial sulcus); 12 = pogonion; 13 = gnathion; 14 = ectocanthion; 15 = alare; 16 = chelion; 17 = gonion; 18 = superaurale; 19 = postaurale; 20 = subaurale; 21 = preaurale; 22 = tragion; 23 = orbitale.

Lateral Points

- **zygion (zy)**–the widest point of the cheek seen in the frontal view.
- **gonion (go)**–the widest point of the lower jaw seen in the frontal view.

Orbital Points

- **ectocanthion (ec)**–the lateral corner (angle) of the eye.
- **endocanthion (en)**–the medial corner (angle) of the eye.
- **iridion laterale (il)**–the most lateral point of the rim of the iris (new term proposal).
- **iridion mediale (im)**–the most medial point on the rim of the iris (new term proposal).

Nasal Points

- **nasion (n)**–in the lateral view, the apex of the frontonasal angle; in the frontal view, the nasion may be plotted on a line tangent to the superior palpebral creases above the upper eyelids (Ashley-Montagu, 1935); with aging, the upper lid tends to fill with fat and fluid and may obscure these creases in which case the nasion frontale will have to be approximated.
- **nasale (na)**–the midpoint of the bony nasal bridge.
- **pronasale (prn)**–the tip of the nose, best seen in the lateral view.
- **subnasale (sn)**–the lowest point of the nose on the midsagittal plane (MSP).
- **alare (al)**–the most lateral point on the "wing" of the nose.

Labial Points

- **superior labial sulcus (sls)**–the point of maximum indentation of the upper lip
- **labiale superius (ls)**–the intersection of the MSP with the vermilion border of the upper lip.
- **stomion (sto)**–the intersection of the MSP with the labial fissure.
- **labiale inferius (li)**–the intersection of the MSP with the vermilion border of the lower lip.
- **chelion (ch)**–the corner of the mouth.

Mental Points

- **labiomentale (lm)**–the intersection of the MSP with the mentolabial sulcus (or inferior labial sulcus).

- **pogonion (pog)**–the most anterior point on the chin seen in the lateral view.
- **gnathion (gn)**–the lowest point of the chin on the MSP.

Auricular Points

- **superaurale (sa)**–the superior pole of the helix.
- **subaurale (sba)**–the inferior pole of the lobe.
- **preaurale (pra)**–the most anterior point of the ear at the base of the tragus.
- **postaurale (pa)**–the most posterior point on the helix.
- **tragion (tr)**–the apex of the tragus.

PLANES

The four most useful facial planes are defined below and illustrated in Figure 1.3.

- **midsagittal plane (MSP)**–divides the face into bilaterally symmetrical right and left sides connecting all midline points from vertex to gnathion (v – gn).
- **midfacial plane (MFP)**–roughly divides the head into upper and lower halves by a horizontal line tangent to the inferior poles of the irises.
- **transverse nasal plane (TNP)**–a horizontal plane parallel to the MFP and passing through the subnasale.
- **transglabellar plane (TGP)**–a horizontal plane passing through the glabella and marking the upper side of the facial square.

AREAS

The areas of the face are, of course, contoured and a true mathematical treatment of its surface would require solid geometry. One of the main objectives of this book, however, is to reveal facial relations via graphic (2-dimensional) facial analysis and thus we must be content with the simpler Euclidian applications of plane geometry. I shall leave the rigors of solid geometry to the growing hoards of computer animators and modelers.

 Facial Geometry

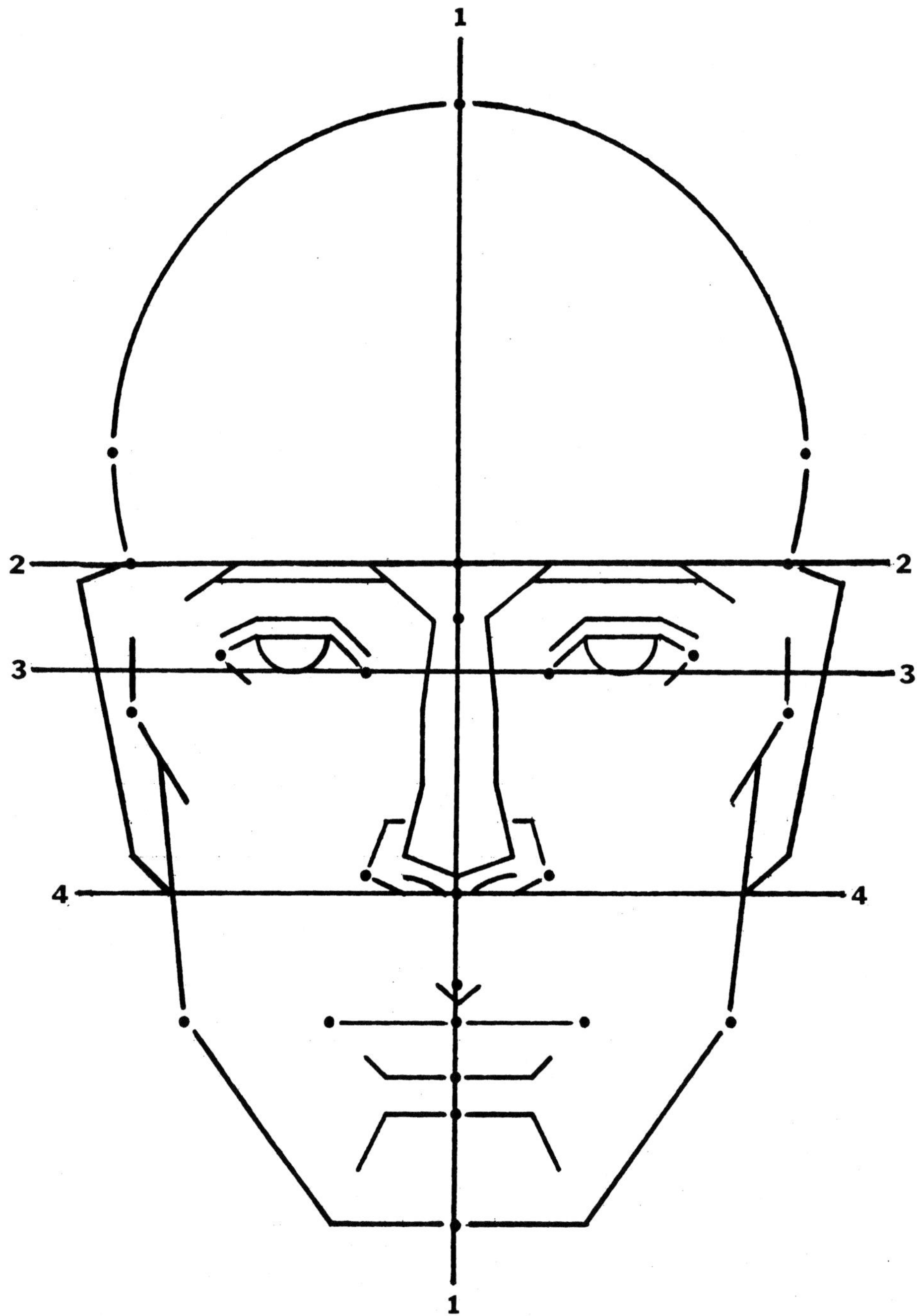

Figure 1.3. Facial Planes: 1 = midsagittal plane (MSP); 2 = transglabellar plane (TGP); 3 = midfacial plane (MFP); 4 = transnasal plane (TNP).

The eight areas of the face as seen in the frontal view (normal frontalis) are demarcated in Figure 1.4. This study is primarily concerned with the relations between the eyes, nose, lips and ear.

1. frontal area
2. nasal area
3. labial area
4. mental area
5. orbital area
6. zygomaxillary area
7. buccomandibular area
8. auricular area

LINES AND GROOVES

In addition to the fixed cephalometric points and planes, there are numerous variable lines and grooves that further define the geography of the face. Faces age at different rates and with different intensities under the influence of stress and gravity. Stress or dynamic lines develop at right angles to the pull of underlying muscles as the skin gradually loses its tonicity. Gravitational effects tend to intensify stress lines converting creases into grooves, folds or pouches. The most common of these surface landmarks are illustrated in Figure 1.5. For a full description of these surface features see George and Singer (1993).

1. **transverse frontal lines**–the forehead wrinkles produced by contraction of the frontalis muscle.
2. **vertical glabellar lines**–these "frown" lines are thought to result from contractions of the paired corrugator supercilii muscles.
3. **transverse nasal line**–one or two horizontal lines at the root of the nose produced by the procerus muscle.
4. **philtrum**–the median depression of the upper lip (described under facial features).
5. **mentolabial groove**–the groove between the lower lip and the chin, probably the result of chin eversion, a hallmark diagnostic feature of *Homo sapiens sapiens.*
6. **mental fovea**–the chin dimple resulting from a gap between the paired mentalis muscles.

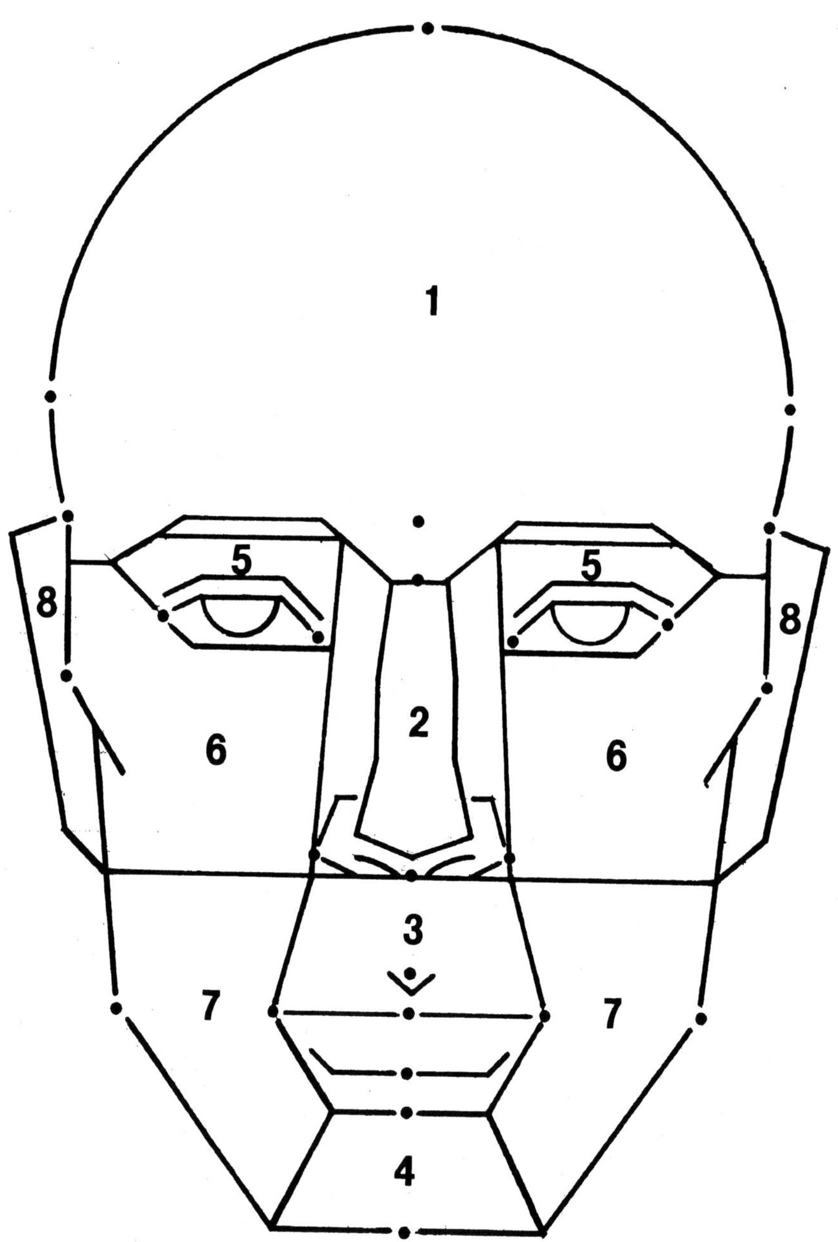

Figure 1.4. Facial Areas: 1 = frontal; 2 = nasal; 3 = labial; 4 = mental; 5 = orbital; 6 = zygo-maxillary; 7 = buccomandibular; 8 = auricular.

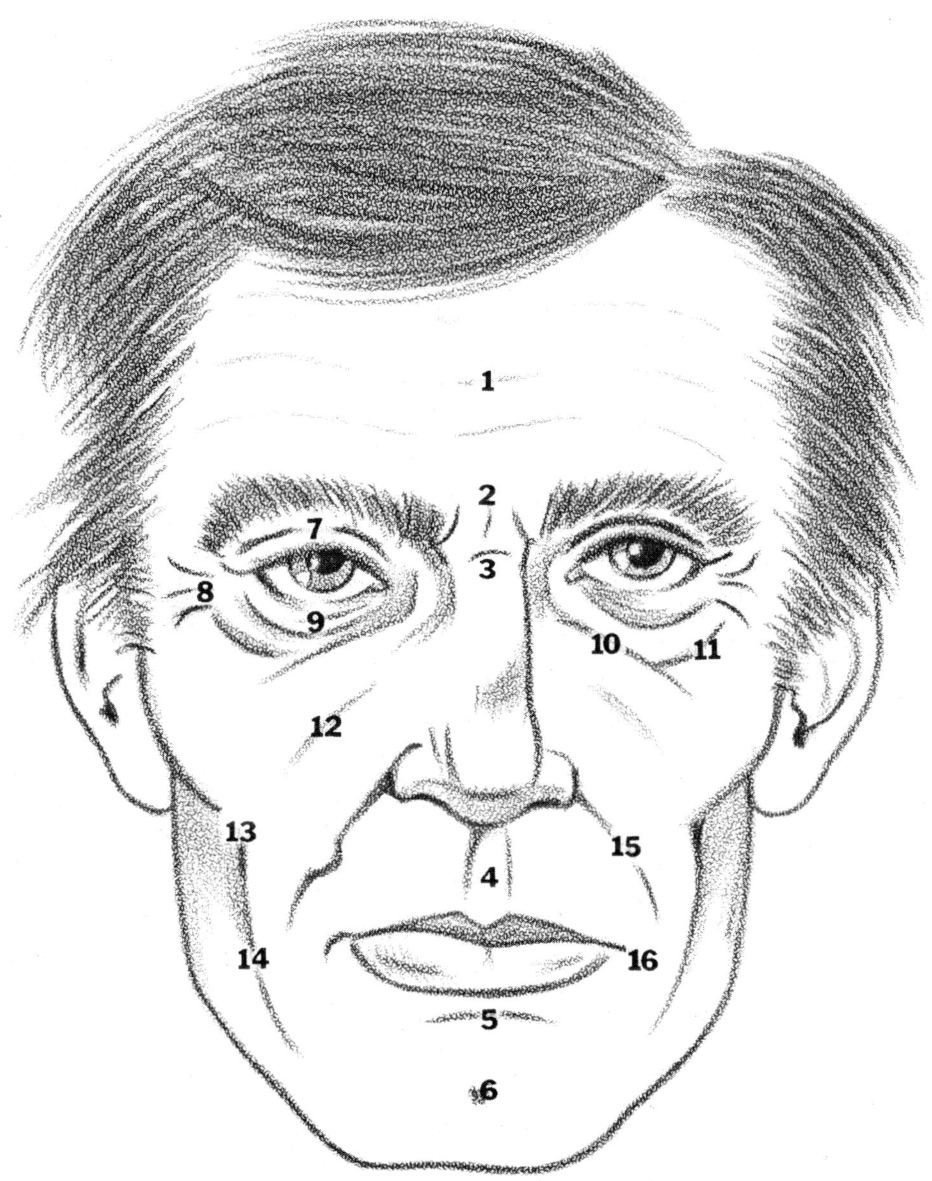

Figure 1.5. Lines and Grooves of the Face: 1 = transverse frontal lines; 2 = vertical glabellar lines; 3 = transverse nasal line; 4 = philtrum; 5 = mentolabial groove; 6 = mental fovea; 7 = superior orbital groove; 8 = lateral orbital lines; 9 = inferior palpebral lines; 10 = inferior orbital groove; 11 = orbitozygomatic line; 12 = maxillary crease; 13 = buccal fovea; 14 = buccomandibular groove; 15 = nasolabial groove; 16 = oromental groove.

7. **superior orbital (or palpebral) groove**–the groove produced by the levator palpebrae superioris muscle when raising the tarsal plate within the orbit, though often obscured by the sagging of the fat-filled palpebral fold, the main target in blepharoplasty.
8. **lateral orbital lines**–the "crow's feet" appearing at right angles to the lateral edge of the orbicularis oculi muscle.
9. **inferior palpebral lines**–variable lines over the lower lid also produced by the orbicularis oculi.
10. **inferior orbital groove**–an oblique groove corresponding to the orbital edge of the maxilla.
11. **orbitozygomatic line**–the lateral continuation of the inferior orbital groove.
12. **maxillary crease**–a highly variable line in the zygomaxillary area.
13. **buccal fovea**–the "cheek dimple."
14. **buccomandibular groove**–a highly variable groove that is more distinctive in males.
15. **nasolabial groove**–a very distinctive groove produced by the pull of the levator labii superioris, levator anguli oris, zygomaticus major and minor, and risorius muscles.
16. **oromental groove**–a variable groove at the labial commissure.

FACIAL FEATURES

In this section, the basic surface features of the eye, nose, mouth and ear will be defined and illustrated in the frontal view (lateral for the ear). These are highly stylized artistic models of Caucasian facial features and are not intended to be composites of the human population.

Orbital Region (Figure 1.6)

1. **superior orbital (palpebral) groove**–the crease between the upper lid and the skin over the brow ridge; a line connecting the right and left grooves passes through the nasion on the MSP; this skin tends to sag with aging, overlapping and obscuring the free margin of the upper lid which should be taken into account when approximating individuals over 50.
2. **free margin of the upper lid *(palpebrae superioris)* with lashes (cilia orbitalis)**–the upper lid is elevated by the levator palpebrae superioris muscle located deep within the orbit; this muscle is

attached to the tarsus, a dense fibrous plate forming the "skeleton of the lid"; this is one of the extraocular muscles and **not** a muscle of facial expression developmentally.

3. **free margin of the lower lid (palpebrae inferioris) with lashes**–this lid also has a tarsal plate but not an independent muscle.
4. **lateral canthus**–the lateral corner of the eye (the cephalometric ectocanthion).
5. **medial canthus**–the medial corner of the eye (the cephalometric endocanthion).
6. **cornea/sclera**–this is the outer fibrous tunic of the eyeball; the sclera is opaque (the white of the eye) but the cornea is that portion in front of the iris that bulges outward and is transparent to light.
7. **iris**–a pigmented disk containing 2 arrangements of smooth muscle which regulates the diameter of the pupil; the dilator pupillae (radial fibers) increases the diameter allowing more light to enter the eye, whereas the constrictor pupillae (circular fibers) decreases the diameter; under normal lighting conditions, the diameter of the pupil is approximately one-third that of the iris.
8. **pupil**–the aperture of the iris in front of the lens.
9. **plica semilunaris**–a fold of the conjunctival mucous membrane at the medial angle of the eye.
10. **lacrimal caruncle**–a small reddish body at the medial angle of the eye containing modified sebaceous and sweat glands.

Nose (Figure 1.7)

1. **root of the nose (nasion)**–the junction between the brow ridge and the nasal bridge.
2. **nasal bridge**–the nasal vault formed by bone proximally and cartilage distally.
3. **supratip break**–the slight indentation near the nasal tip formed by the alar cartilage overlapping the septal and lateral nasal cartilages.
4. **nasal tip**–the most anterior point of the nose (pronasale).
5. **columella**–the flexible underpart of the nasal tip partitioning the nostrils and formed by the medial crura of the alar cartilages.
6. **wing of the nose**–this is formed by the flaring of the lateral crus of the alar cartilage.
7. **alar groove**–the crease between the wing of the nose and the cheek.

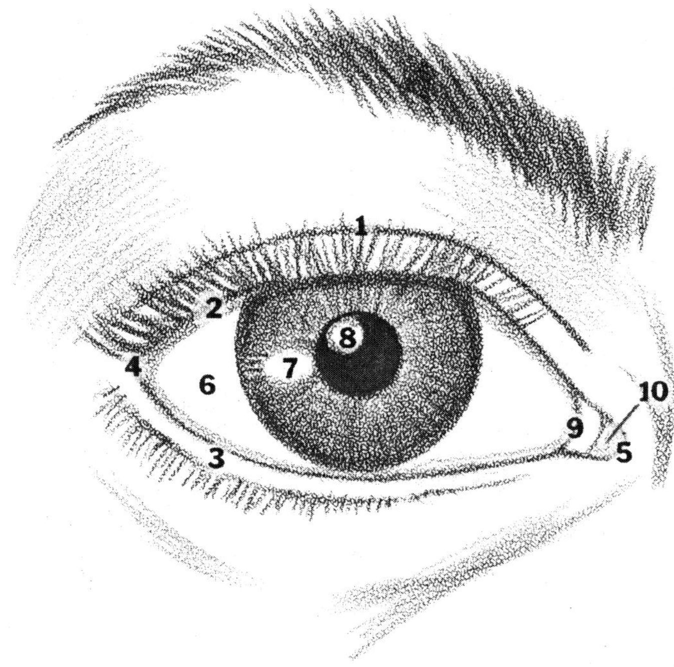

Figure 1.6. Orbital Region: 1 = superior orbital groove; 2 = free margin of the upper lid; 3 = free margin of the lower lid; 4 = lateral canthus; 5 = medial canthus; 6 = sclera; 7 = iris; 8 = pupil; 9 = plica semilunaris; 10 = lacrimal caruncle.

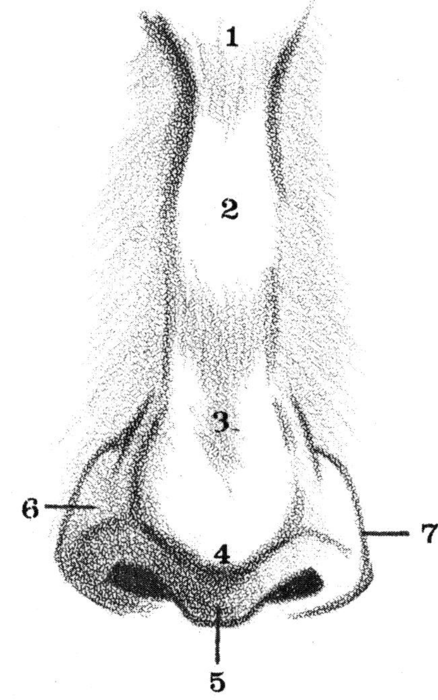

Figure 1.7. Nose: 1 = nasion; 2 = nasal bridge; 3 = supratip break; 4 = nasal tip; 5 = columella; 6 = nasal wing; 7 = alar groove.

Lips (Figure 1.8)

Both lips are formed by an outer "skin" (keratinized epidermis) and an inner "mucous membrane" (unkeratinized epithelium). The red vermilion area represents the transitional zone where the translucent prekeratin allows the capillary blood to show through.

1. **philtrum**–the median depression in the upper lip (from the Greek for "loving").
2. **upper vermilion border**–the demarcation in the upper lip between skin and mucous membrane; this border is sometimes referred to as Cupid's Bow due to its sinuous shape resembling the little Love God's archery equipment (this may also explain the origin of the term "philtrum").
3. **lower vermilion border**–the less romantic edge of the lower lip.
4. **labial fissure**–the slit between the lips.
5. **tubercle**–the median protuberance of the upper lip.
6. **alae**–the paired wings of the upper lip on either side of the tubercle.
7. **labial commissure**–the corner (angle) of the mouth more commonly referred to as the cephalometric chelion (ch).
8. **median sulcus**–the slight midline groove in the lower lip.
9. **tori**–the paired swellings forming the body of the lower lip.
10. **mentolabial sulcus**–the groove between the lower lip and the top of the mental protuberance (chin).

Ear (Figure 1.9)

1. **helix**–the outer rim of the auricle (external ear).
2. **Darwin's tubercle**–a variable protuberance on the upper portion of the helix.
3. **lobe**–the fleshy, non-cartilaginous part of the auricle at its inferior pole.
4. **helical groove**–the groove between the helix and the antihelix.
5. **antihelix**–the inner "Y"-shaped ridge in the middle of the auricle.
6. **superior crus**–the upper limb of the antihelix.
7. **triangular fossa**–the depression between the limbs (crura) of the antihelix.
8. **inferior crus**–the lower limb of the antihelix.
9. **cymba**–the depression in the conchal fossa above the inturned crus of the helix.

10. **cavum**–the depression in the conchal fossa below the crus of the helix.
11. **tragus**–a tongue-like projection of cartilage in front of the opening of the external auditory meatus ("tragi" = hairs).
12. **antitragus**–a cartilaginous bump at the inferior end of the antihelix opposite the tragus.
13. **tragal notch**–the cleft between the tragus and the antitragus.
14. **tragion**–the cephalometric point at the helicotragal junction which corresponds with the craniometric porion.

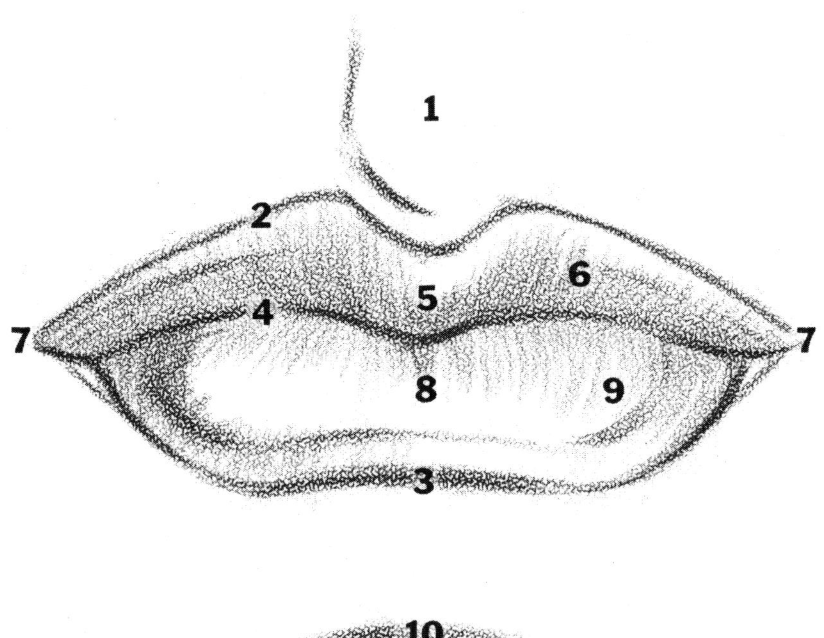

Figure 1.8. Lip: 1 = philtrum; 2 = upper vermilion border; 3 = lower vermilion border; 4 = labial fissure; 5 = tubercle; 6 = ala; 7 = labial commissure; 8 = median sulcus; 9 = torus; 10 = mentolabial sulcus.

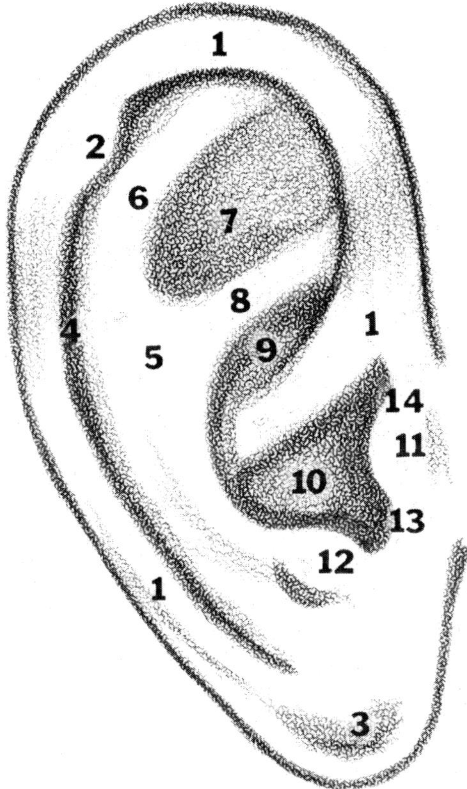

Figure 1.9. Ear: 1 = helix; 2 = Darwin's tubercle; 3 = lobe; 4 = helical groove; 5 = antihelix; 6 = superior crus; 7 = triangular fossa; 8 = inferior crus; 9 = cymba; 10 = cavum; 11 = tragus; 12 = antitragus; 13 = tragal notch; 14 = tragion.

Chapter 2

FACIAL GEOMETRY

THE GOLDEN SECTION

(Figure 2.1)

During the Golden Age of Pericles (approximately 460–430 B.C.), the Athenians were a rollicking crowd busily cultivating what was to become known as Western civilization. Their contributions enveloped and invigorated philosophy, all of the arts and sciences both physical and social, mathematics, democratic politics, economics, education, Olympian athletics – and warfare. In their free moments they were especially fond of games, mathematical posers in particular. One such entertainment may be stated as follows:

Where can line AB be divided at point C such that segment CB:AC = AC:AB?

A________________________C____________B
                         ?

The solution given in the following steps was first formally recorded by Euclid, the Father of modern geometry:

1. Draw line AB
2. Erect segment BD perpendicular to AB, where BD = 1/2 AB.
3. Draw the hypotenuse AD
4. From point D with radius = DB, draw an arc that intersects AD at point E
5. From point A with radius = AE, draw an arc that intersects AB at point C
6. Thus AC = the Golden Section = 0.618

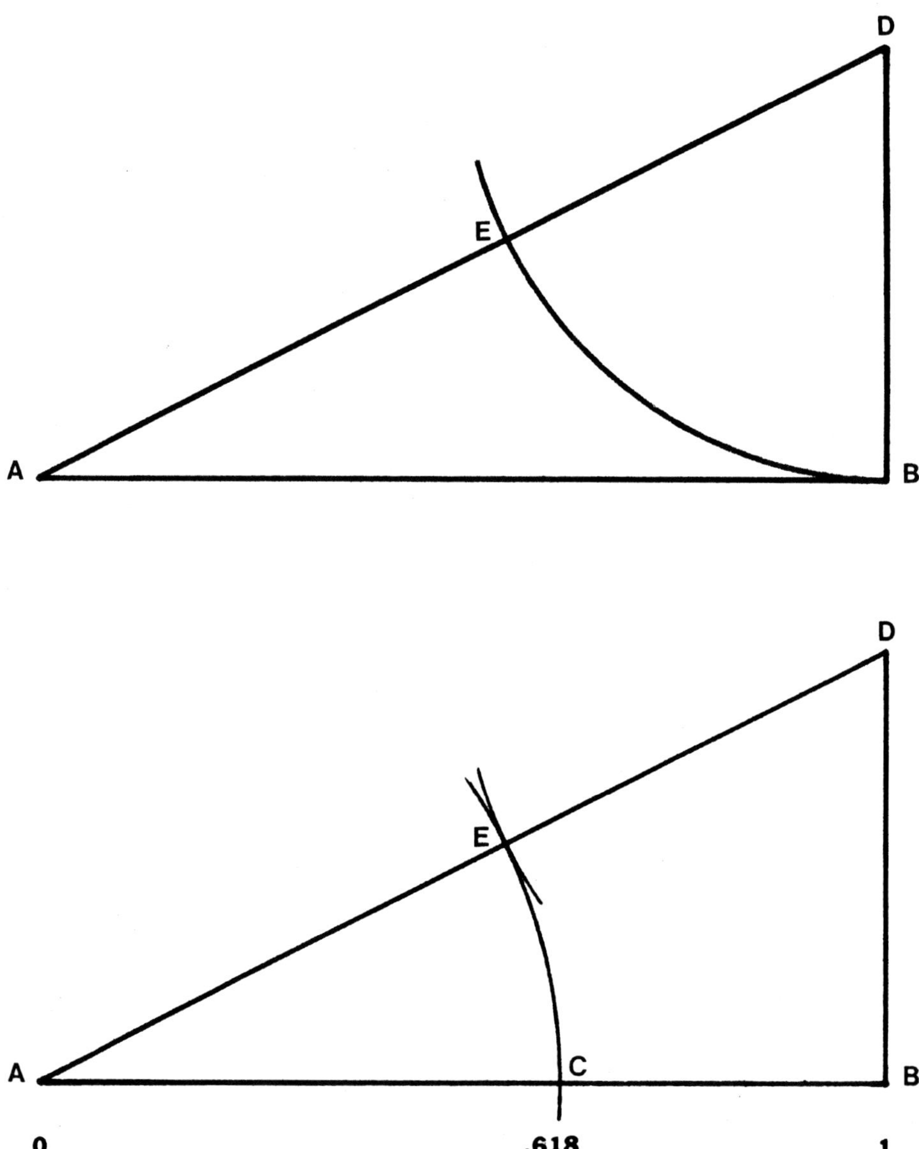

Figure 2.1. The Golden Section.

This remarkable number (0.618) is the only one in mathematics which when divided into unity (1.0) yields its own reciprocal (1.618)! It is not known who first solved this puzzle or even if it was rationally deduced or just a happy accident of compass doodling. Whatever, it has long been known as the golden number, golden cut, golden section, golden proportion or golden ratio and was popularized by Luca Pacioli in 1509 in his classic text, *De Divina Proportione* which was illustrated by no lesser a light than Leonardo Da Vinci. Da Vinci's "Proportion Man" must be the most famous logo in all of art history. For scholarly treatments of the Divine Proportion, the interested reader is referred to the excellent books by Huntley (1970) and Livio (2002).

THE GOLDEN RECTANGLE

(Figure 2.2)

The Greeks, however, had their own share of artistic luminaries. Pliny considered the Greek sculptor Myron of "Discobolus" fame to be the first to achieve realism in the depiction of the human figure. Polyclitus deserves special mention since it was his life's work to establish a canon or rule for the correct proportion of every part of the human body. He has been called the "Pythagoras" of sculpture for seeking a divine mathematics of symmetry and form. But perhaps the most outstanding of all the Greek sculptors was Phidias who directed the construction of the Parthenon, designed its famous friezes–the Elgin Marbles and created its principle inhabitant, the 38′ gold and ivory statue of the Athena Parthenos. He later outdid himself with the 42′ stature of a seated Zeus for the temple at Olympia. Phidias was so enamored of the golden section, considering the 3:5 ratio to be the ultimate expression of mathematical beauty, much prettier than 1:2 or 2:3 or 3:4 that he used it repeatedly in his sculptural and architectural works. Indeed, it is the master plan of the Parthenon itself. Such was his influence that the golden section is referred to mathematically as Phi.

It should now be apparent that a golden rectangle is nothing more glamorous than a 3 × 5 index card. To satisfy mathematical purists, the steps for constructing a golden rectangle are given below:

1. Draw square ABCD
2. Bisect line AB at E

3. Draw line EC
4. Using radius EC, swing an arc upward until it intersects extended line AB at point F
5. Drop a perpendicular from point F until it intersects extended line DC at point G
6. The short and long sides of AFGD are in a golden ratio = 0.618 (if constructed with drafting accuracy).
7. Rectangle BFGC is also golden and such golden sectioning is infinite!

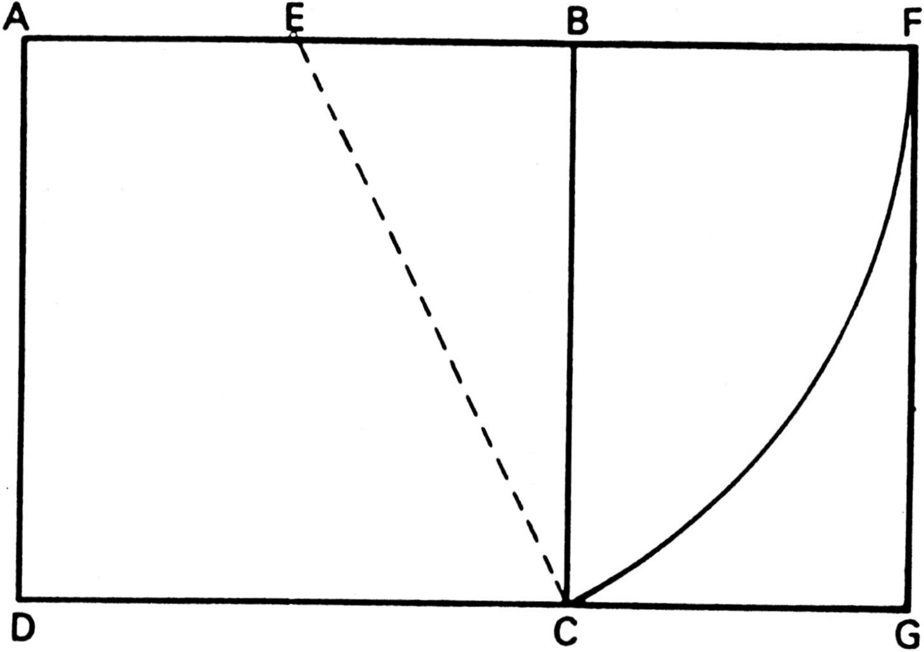

Figure 2.2. Construction of the Golden Rectangle.

It would not be proper to leave the golden number without reference to the famous Fibonacci series. Leonardo Fibonacci a.k.a. Leonardo of Pisa (c. 1170–1250 A.D.), an Italian numbersmith who introduced Hindu-Arabic numbers to Europe, achieved mathematical immortality by his ability to add 1 + 1, then 1 + 2, then 2 + 3, 3 + 5 and 5 + 8 and so on. After about ten such successive additions, each new addition is precisely 1.618 times the previous number and this ratio remains constant from here to infinity (Ricketts, 1982). This sequence has been intensely studied, spawning its own cult of mathematicians ("Phi-bonatics") who spread the gospel via such journals as *The Fibonacci Quarterly,* books and myriad web sites (www.golden mean . . .). The golden number is only one of many irrational constants underlying the very mysterious nature of the universe such as Archimedes' constant (pi = 3.1416), Pythagoras' constant (square root of 2 = 1.4142), and the speed of light (186,291 mi/sec). Other than being intellectually intoxicating, the Fibonacci series has little application to forensic art.

PROPERTIES OF THE GOLDEN RECTANGLE

(Figure 2.3)

To illustrate the properties of a golden rectangle, a figure was constructed on graph paper measuring 10 squares (units) per inch with sides = 36 × 60 units.

OBSERVATIONS

- line FL divides the rectangle into right and left sides.
- line CI divides the rectangle into upper and lower halves.
- the sides of the square ADHK = 36 units.
- line BJ divides the square into upper and lower halves.
- line BJ and the diagonals DK and HA intersect line FL at point X.
- a "golden isosceles triangle" may be drawn within the rectangle by connecting points A-F-K (note: a true golden triangle will have base angles of 72° and a vertex angle of 36° but since this is a "rounded off" golden rectangle, the golden triangle has angles closer to 73° · 34° · 73°).

THE GOLDEN FACE

(Figure 2.4)

We can now draw a human face within the golden rectangle and substitute cephalometric points for the geometric points in Figure 2.3.

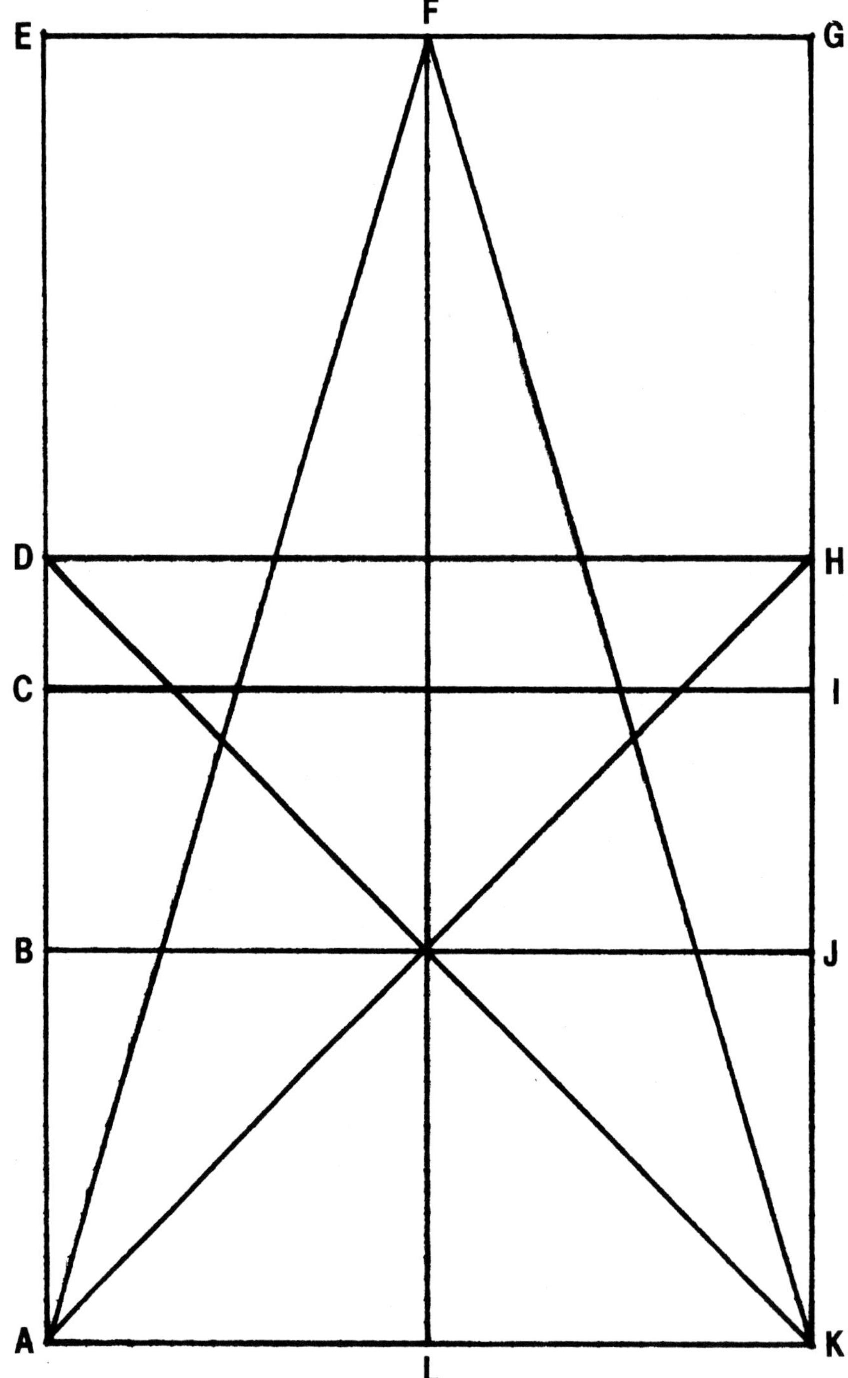

Figure 2.3. Properties of the Golden Rectangle.

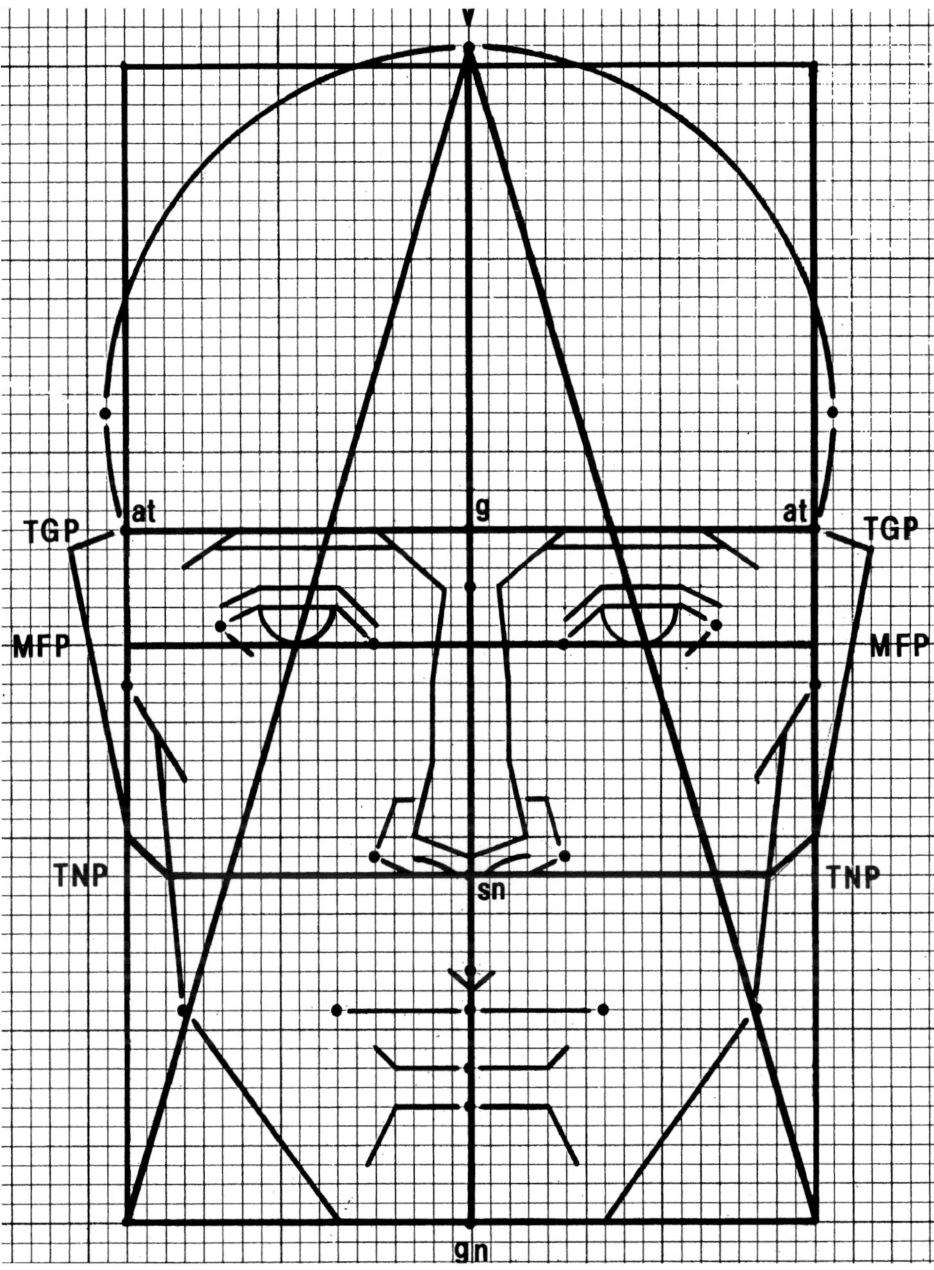

Figure 2.4. The Golden Face: v = vertex; g = glabella; sn = subnasale; gn = gnathion; at = auriculotemporale; TGP = transglabella plane; MFP = midfacial plane; TNP = transnasal plane.

OBSERVATIONS

- the midsagittal plane (v-gn) divides the head into right and left halves.
- the midfacial plane (MFP) divides the head into upper and lower halves and is tangent to the inferior poles of the irises.
- the transglabellar plane (TGP) connects the auriculotemporal points and passes through the glabella (Note: the line at-at forms the top side of the facial square and = 36 units).
- the transnasal plane (TNP) divides the facial square into upper and lower halves and passes through the subnasale.
- moving down 18 units from point v, a circle can be inscribed around the golden rectangle representing the braincase (note: in most individuals the braincase or neurocranium is wider than the face or viscerocranium and thus slightly exceeds the boundaries of the golden rectangle).
- the legs of a golden triangle constructed within a golden rectangle pass through the centers of the pupils!

NOTE 1: It must be made clear that the "geometric face" in Figures 2.4 to 2.10 was drawn to fit the geometric parameters under discussion. It is by no means intended to represent the ideal "golden" face of legend. Many have attempted the synthesis of the ideal (= most beautiful) face based on some mathematical formula of harmonious perfection but in the final analysis this remains a judgement based on individual preferences rather than biomathematical law. Over the past few years, I have contrived many different faces which "fit" different sets of circles, rectangles and triangles, all smacking of Divine intervention. The face presented here is just another variation of the human facial theme. It is not intended to be a revelation.

NOTE 2: To avoid the mathematical tedium of dealing with hundredths and thousandths of a unit, Phi (0.618) has been rounded off to a manageable 0.60 and given a range of 0.56–0.64. Thus, some of the figures illustrated in this chapter may be slightly less than 24 karat. Forensic artists deal with approximations and need not be encumbered with engineering precision. The Phi values for all rectangles in this study were obtained by dividing the short side of the rectangle by its long side.

THE FACIAL SQUARE

(Figure 2.5)

OBSERVATIONS

- the lines of the facial square pass through g (superiorly), gn (inferiorly) and zy bilaterally; the upper corners of the square meet at the hypothetical auriculotemporal points (at) (note: the upper poles of the ears are not fused to the sides of the head and the actual contact point is much lower).
- the diagonals pass through ec, sn and ch.
- sn should thus be considered as the center of the face with great potential for predicting ec and ch in forensic approximations.
- the diagonals of the facial square generate four isosceles right triangles ($45° \cdot 90° \cdot 45°$).

TWO × THREE RECTANGLES

(Figure 2.6)

A seemingly endless number of facial rectangles have been proposed by numerous authors. In this analysis only those rectangles that connect critical cephalometric points are evaluated. Not all are golden, i.e., $3 \div 5 = 0.6$, since the rectangles illustrated in Figure 2.6 are closer to $2 \div 3$ giving a ratio of 0.667.

OBSERVATIONS

- a 2×3 rectangle hits the following points:
  1. the superior horizontal line passes through n.
  2. the inferior horizontal line passes through gn.
  3. the vertical sides are tangent to the lateral edges of the irises(il).

This is not a very significant rectangle since it does not relate the eyes to either the nose or mouth.

- another 2×3 rectangle hits the following points:
  1. the superior horizontal line passes through n.
  2. the inferior horizontal line passes through sn.

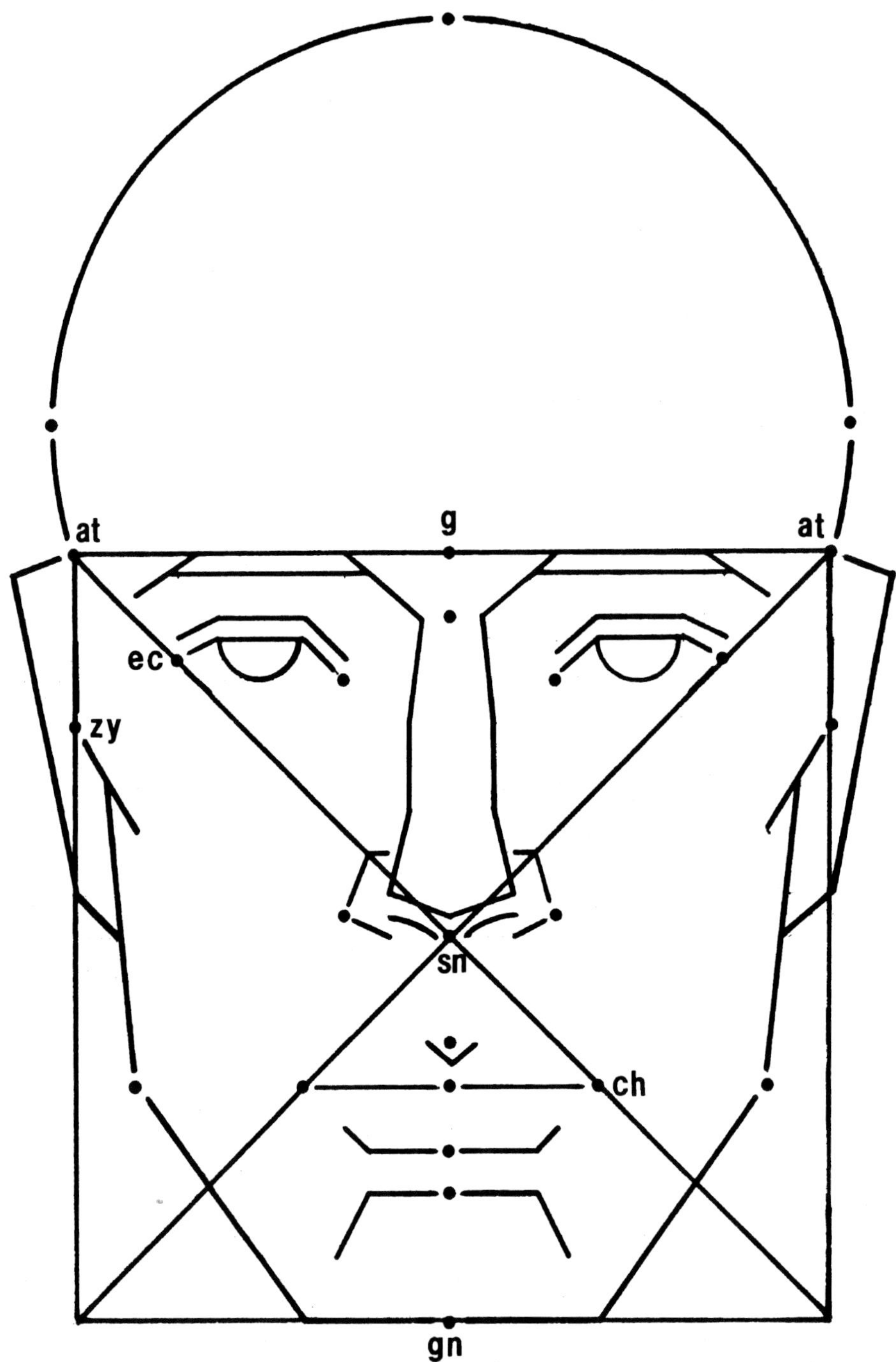

Figure 2.5. The Facial Square: g = glabella; sn = subnasale; gn = gnathion; at = auriculotemporale; ec = ectocanthion; zy = zygion; ch = chelion.

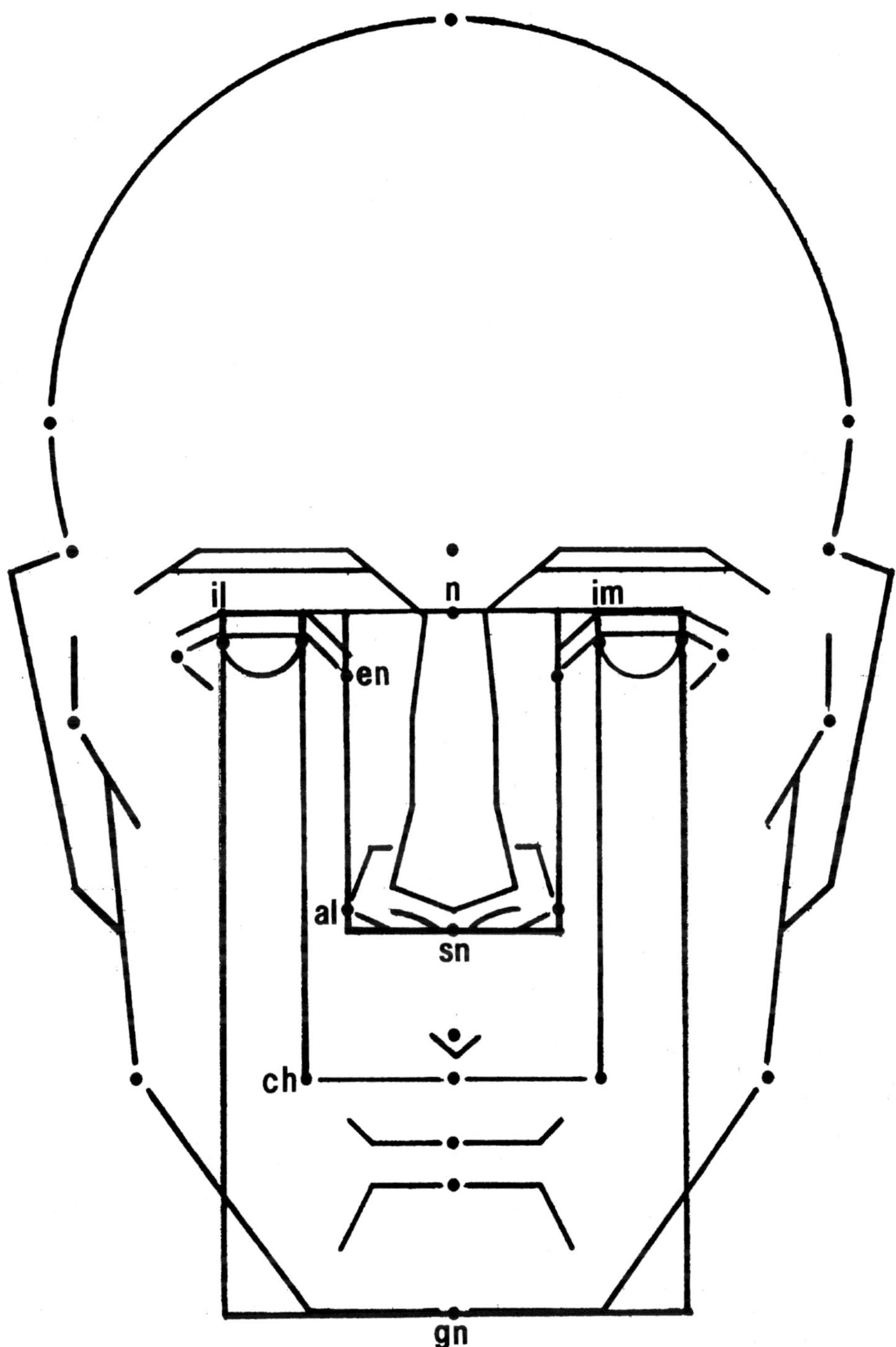

Figure 2.6. Two x Three Rectangles: n = nasion; sn = subnasale; gn = gnathion; en = endocanthion; al = alare; ch = chelion; im = iridion mediale; il = iridion laterale.

3. the vertical sides connect the medial canthi (en) with the alae (al) of the nose.

This is a critical rectangle since it relates the inner corners of the eyes to the sides of the nose.

• another rectangle can be constructed with the following contact points:
  1. the superior horizontal line passes through n.
  2. the inferior horizontal line is the interchelial line (ch-ch).
  3. the vertical sides connect the medial edges of the irises (im) with the cheilia (ch).

This is a significant rectangle since it relates the eyes to the mouth but it misses being a 2×3 rectangle by one unit (14/22 instead of 14/21). With a quotient of 0.636, it is within the golden range.

THE FACIAL TRIANGLES

(Figure 2.7)

The equilateral facial triangle (ec-li-ec) has long been known to artists since it remains equilateral in both relaxed and smiling faces (Hamm, 1982). When smiling, the upper lip is pulled superiorly by the levator labii superioris complex while the lower lip is merely stretched by the risorius. This is a very practical triangle since it relates the eyes to the mouth and can be used to approximate the position of the elusive lateral canthus (ec) in facial approximations (George, 1993; Taylor, 2001).

THE NASAL TRIANGLES

(Figure 2.8)

A golden triangle (ch-n-ch) frames the nose with legs just medial to the alar points. A larger triangle (ch-g-ch) hits the alar points but the apical angle is less than golden (32° rather than 36°). These triangles are extremely helpful in forensic facial approximations, providing guidelines for assessing nasal width.

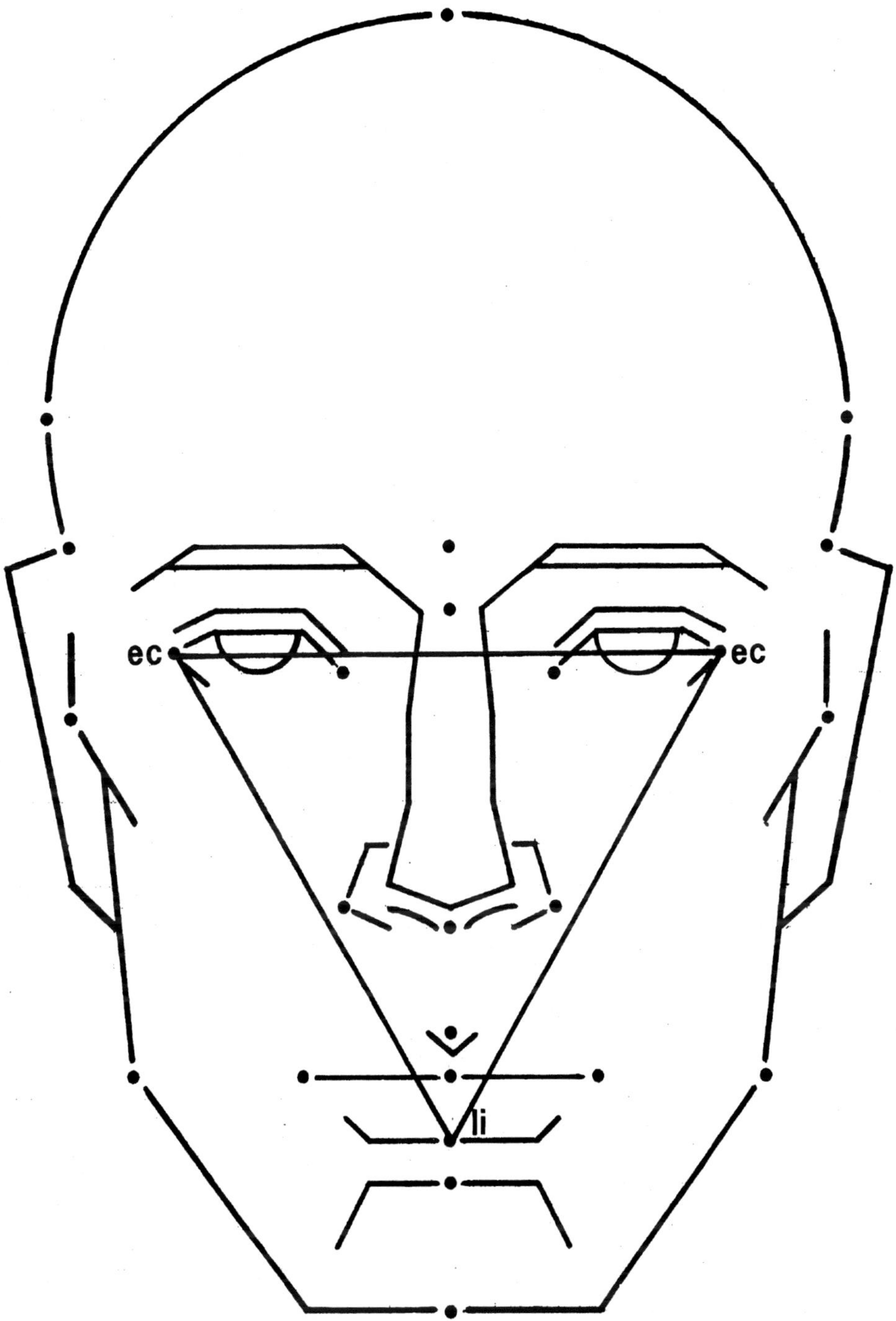

Figure 2.7. The Facial Triangle: ec = ectocanthion; li = labiale inferius.

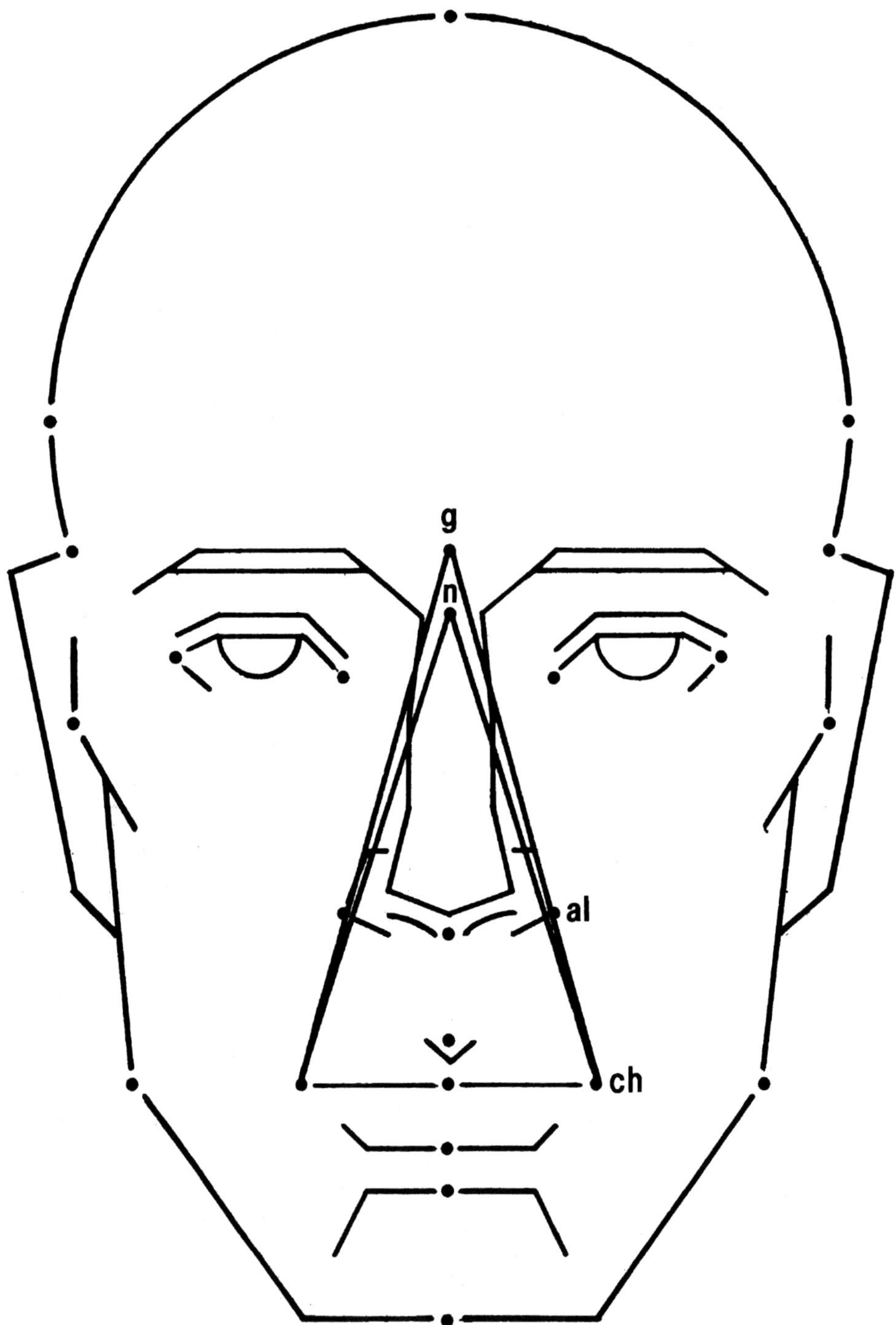

Figure 2.8. The Nasal Triangles: g = glabella; n = nasion; al = alare; ch = chelion.

NEARLY GOLDEN TRIANGLES

(Figure 2.9)

As illustrated in Figure 2.4, a golden triangle can be inscribed within the golden facial rectangle with the apex at the vertex (v) and legs passing through the pupils. A true golden triangle will have angles of 72° · 36° · 72° but since we have rounded off the golden mean from 0.618 to 0.600, the triangle is slightly tarnished with the apical angle being closer to 34°. Three other nearly golden triangles also connect critical facial points using a baseline that passes through the nasion (n) and paired superior palpebral (orbital) grooves.

OBSERVATIONS

- a nearly golden triangle connects the lateral edges of the irises (il) with the apical gnathion (gn).
- a nearly golden triangle connects the centers of the pupils with the apical labiomentale (lm)—note: the legs also pass very close to the alar points.
- a nearly golden triangle connects the medial edges of the irises (im) with the apical stomion.

THE FACIAL TRAPEZOID

(Figure 2.10)

OBSERVATIONS

- an isosceles trapezoid can be constructed with the following contact points:
  1. the superior baseline passes through n and the paired superior palpebral grooves.
  2. the inferior baseline is the interchelial line (ch-sto-ch).
  3. the legs connect ch with ec and extend to the superior baseline.

The facial trapezoid can be trisected into 3 nearly golden triangles: a median (ch-n-ch), the legs of which also pass close to al and paired laterals (ec-ch-n) with apical angles in each close to 36°.

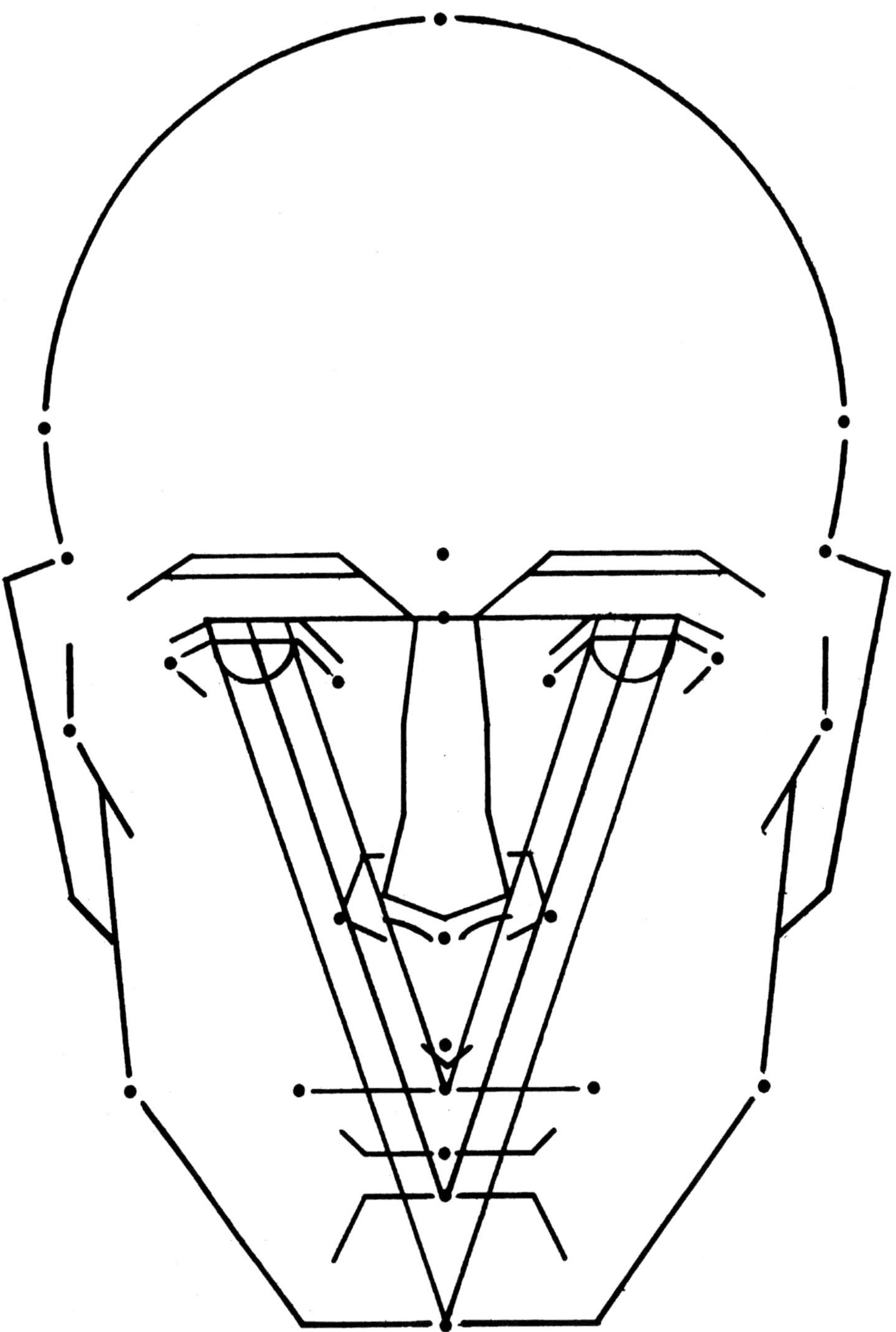

Figure 2.9. Nearly Golden Triangles.

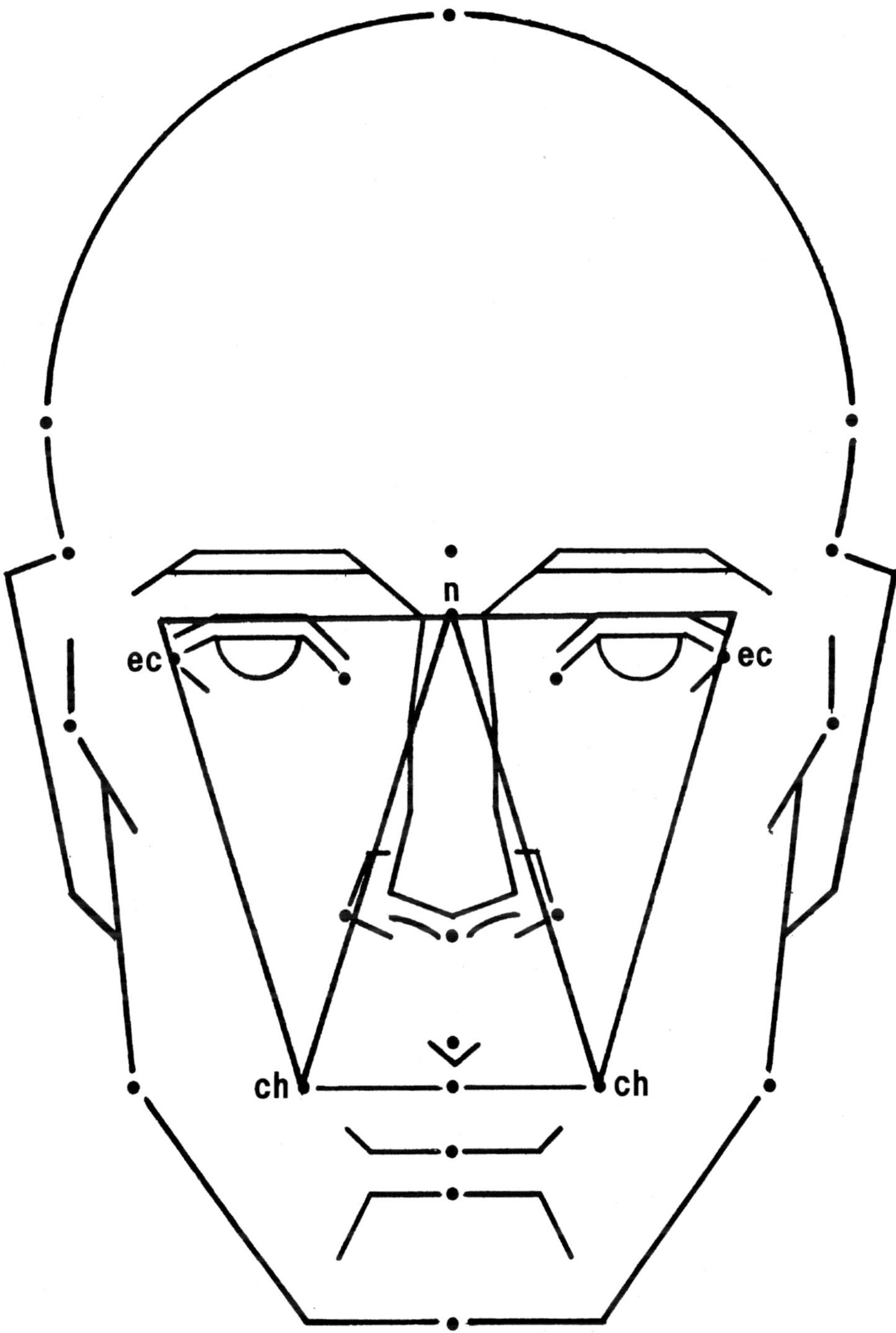

Figure 2.10. The Facial Trapezoid: n = nasion; ec = ectocanthion; ch = chelion.

Chapter 3

GRAPHIC FACIAL ANALYSIS:
THE FRONTAL VIEW

Graphic facial analysis is the essence of forensic art. Whether rendering a composite sketch or "resur-facing" a skull, the forensic artist must be able to accurately assess facial relations and proportions. Most forensic artists start composite sketches with a template based on the classical artistic canons of facial proportions (George, 1993) which is then adjusted as the witness description unfolds. This process will be greatly enhanced if the artist is familiar with the dozen or so basic indices that interrelate facial features.

Mathematically, an index is simply a ratio in which the numerator is divided by the denominator and the quotient multiplied by 100 to express the result as a whole number. As an index, Phi would be 61.8. In their definitive study, Farkas and Munro (1987) statistically evaluated 166 facial proportions in children and young adults. In this more modest study, we will examine the indices and angles that have the most utility for forensic artists. The means and ranges cited have been largely derived from the aforementioned work for 18-year-old Caucasian males and females and from the author's unpublished data on adult males and females taken from frontal portraits in the photographic atlas *Heads* by Alex Kayser (1985).

The following illustrations of the various facial indices were drawn on the "golden face" depicted in Chapter 2. A graphic facial analysis of this figure shows that the golden face (a.k.a. "Phidias") deviates from the normal range in several facial relations.

37

FACIAL INDICES

FACIAL LENGTH INDEX

(n-gn/zy-zy × 100)

(Figure 3.1)

This index measures the length of the face relative to its width.

	Wide Face	Normal Range	Narrow Face
Males	← 83.30	83.40–93.60	93.70 →
Females	← 81.40	81.50–90.86	90.97 →

This is the most well-known of the facial indices. In anthropometry, a wide face is said to be euryprosopic, a medium face is mesoprosopic and a long face is leptoprosopic. When all four points are connected, a facial "kite" is created. This is almost an optical illusion in that it "seems" as if line n-gn is longer than line zy-zy but this is only so in faces that are hyperleptoprosopic. It is also interesting to note that the upper sides (n-zy) of the quadrilateral kite pass near the lower poles of the irises and the lower sides (zy-gn) pass through the chelia!

MANDIBULO-FACIAL INDEX

(go-go/zy-zy × 100)

(Figure 3.2)

This index measures the width of the jaw relative to the width of the face.

	Narrow Jaw	Normal Range	Wide Jaw
Males	← 66.90	67.00–74.60	74.70 →
Females	← 65.80	65.90–74.30	74.40 →

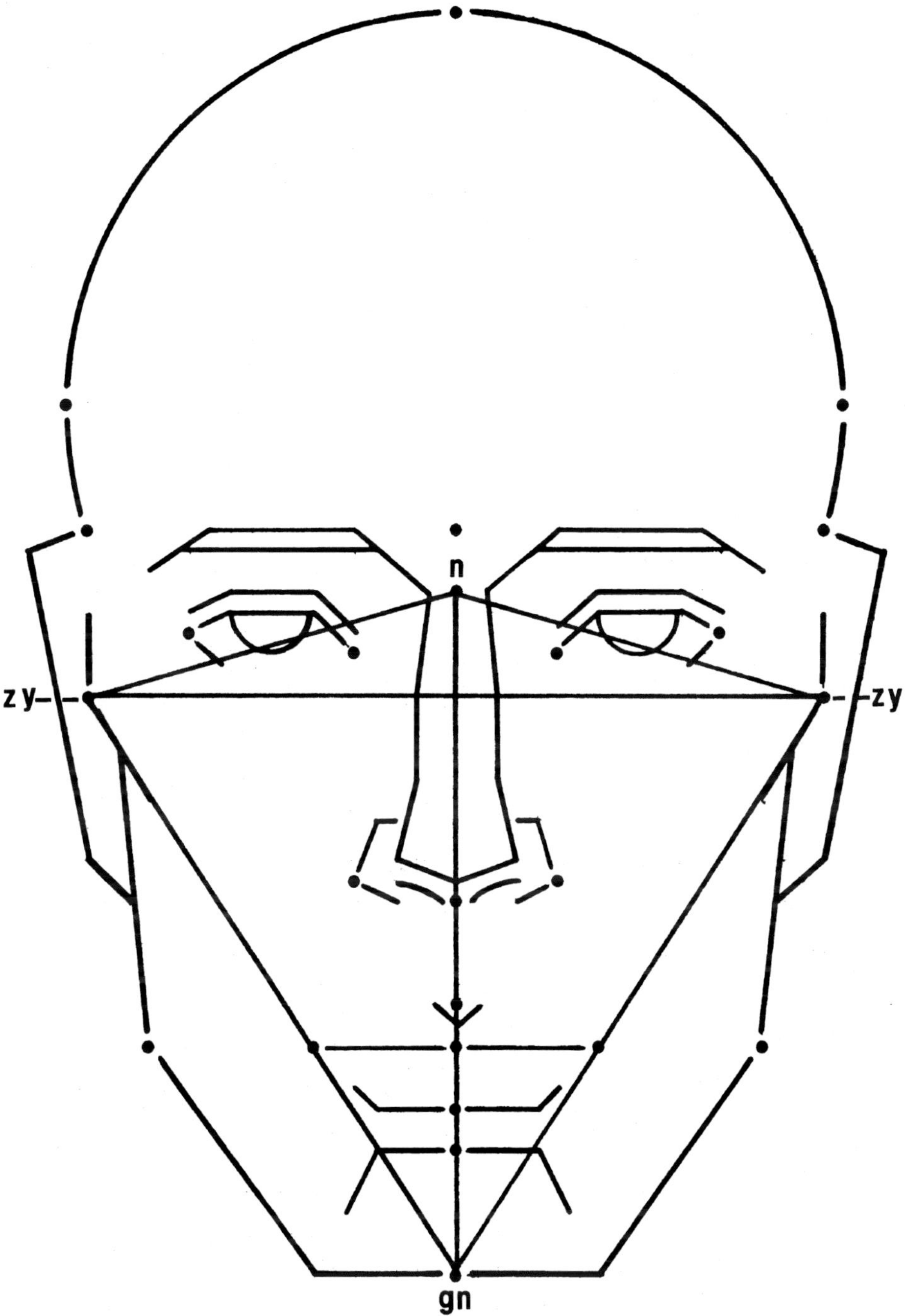

Figure 3.1. Facial Length Index: n = nasion; gn = gnathion; zy = zygion.

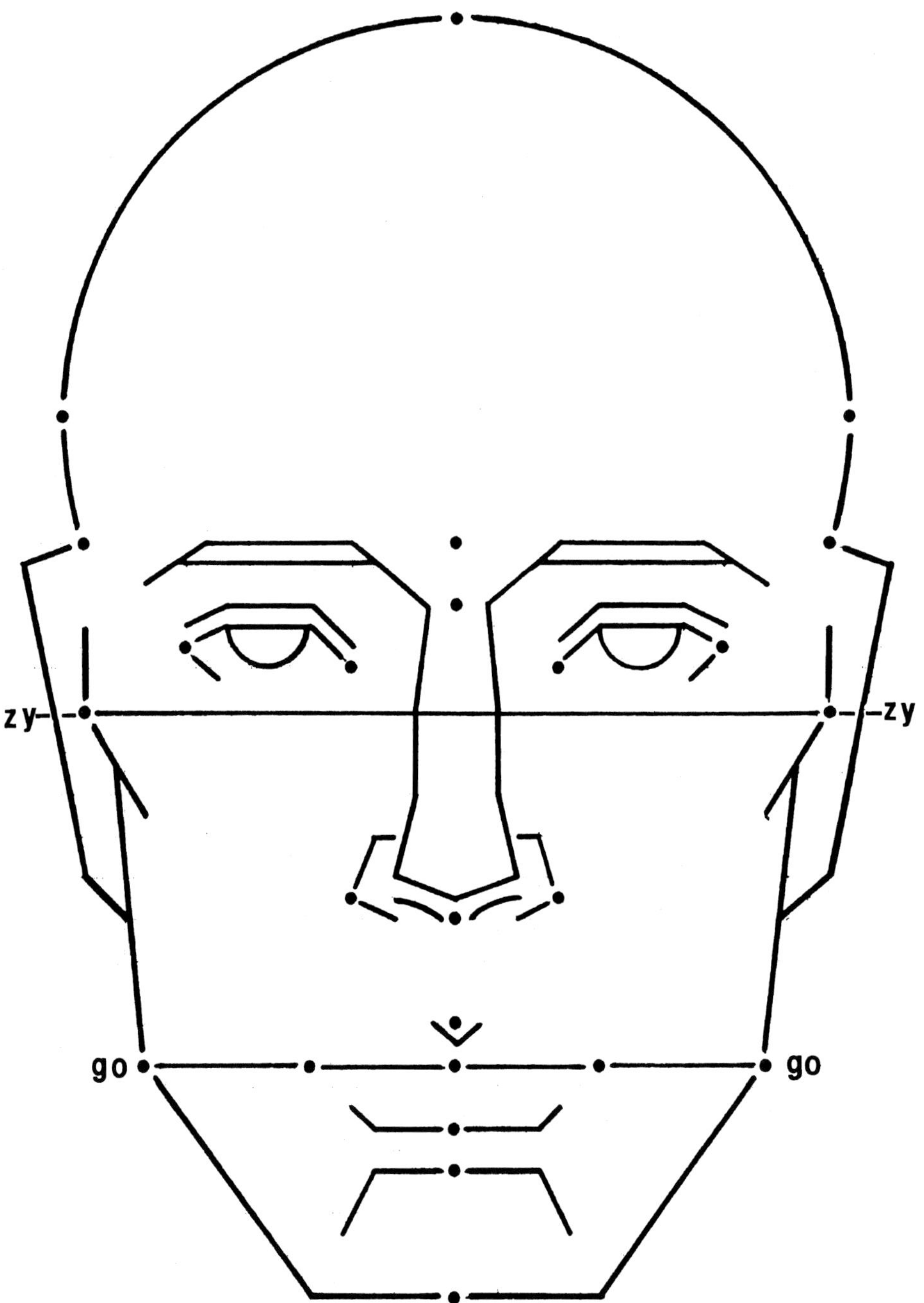

Figure 3.2. Mandibulo-Facial Width Index: zy = zygion; go = gonion.

The facial length and mandibular width indices determine the overall shape of the face. The wider the mandibular index, the squarer the face and when narrow the face is more triangular.

INTERCANTHAL INDEX

(en-en/ec-ec × 100)

(Figure 3.3)

This index measures the distance between the eyes relative to ocular width.

	Eyes Close Set	Normal Range	Eyes Far Apart
Males	← 34.60	34.70–38.90	39.00 →
Females	← 34.30	34.40–38.40	38.50 →

Farkas et al. (1985) found the interocular width (en-en) to be equal to the width of the eye (ec-en), the classical canon, in 33 percent of cases (n = 103). The intercanthal width was narrower in 15.5 percent and wider in 51.5 percent.

NASAL LENGTH INDEX

(n-sn/n-gn)

(Figure 3.4)

This index measures the length of the nose relative to facial length.

	Short Nose	Normal Range	Long Nose
Males	← 41.00	41.10–46.30	46.40 →
Females	← 41.50	41.60–46.00	46.10 →

 Facial Geometry

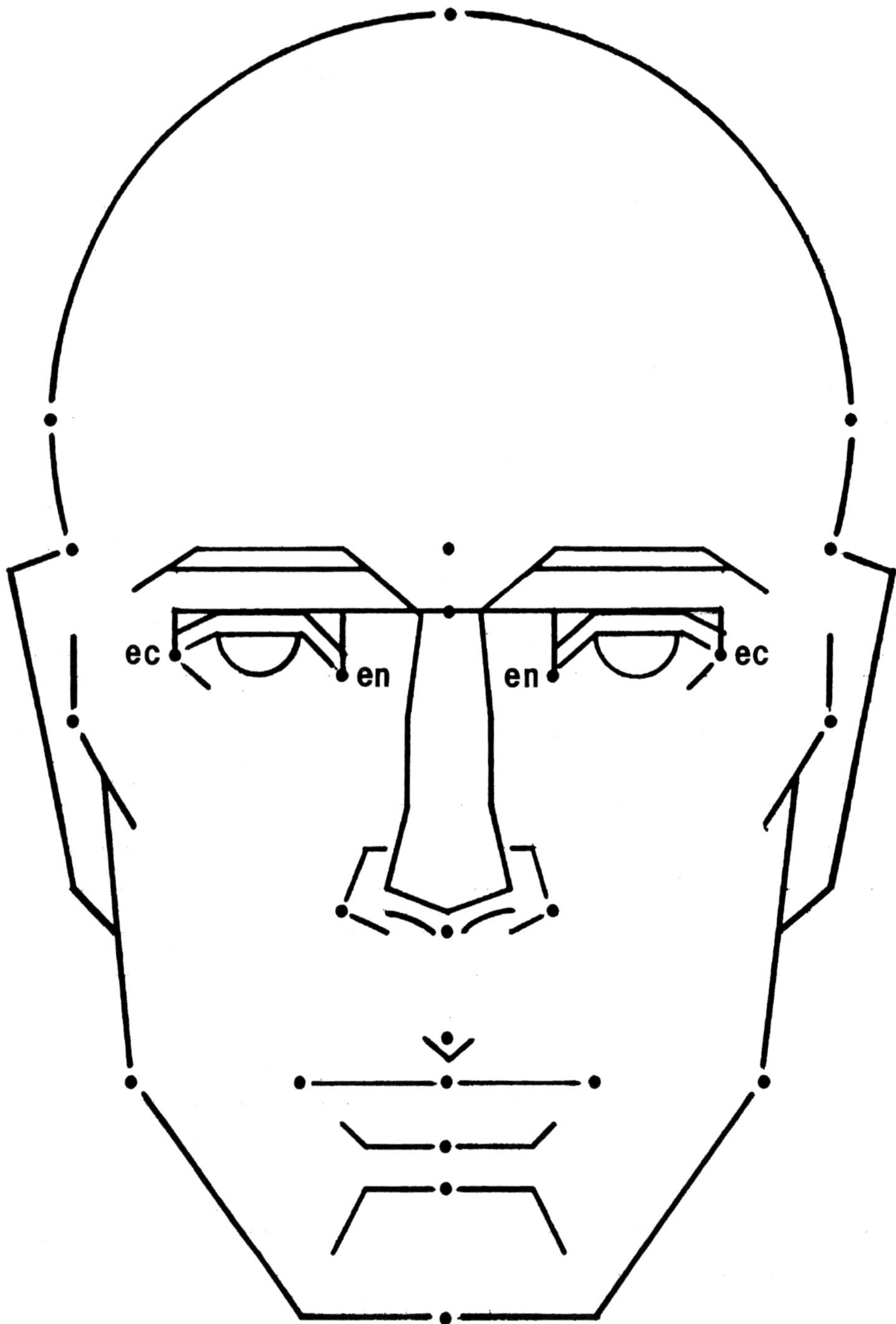

Figure 3.3. Intercanthal Index: en = endocanthion; ec = ectocanthion.

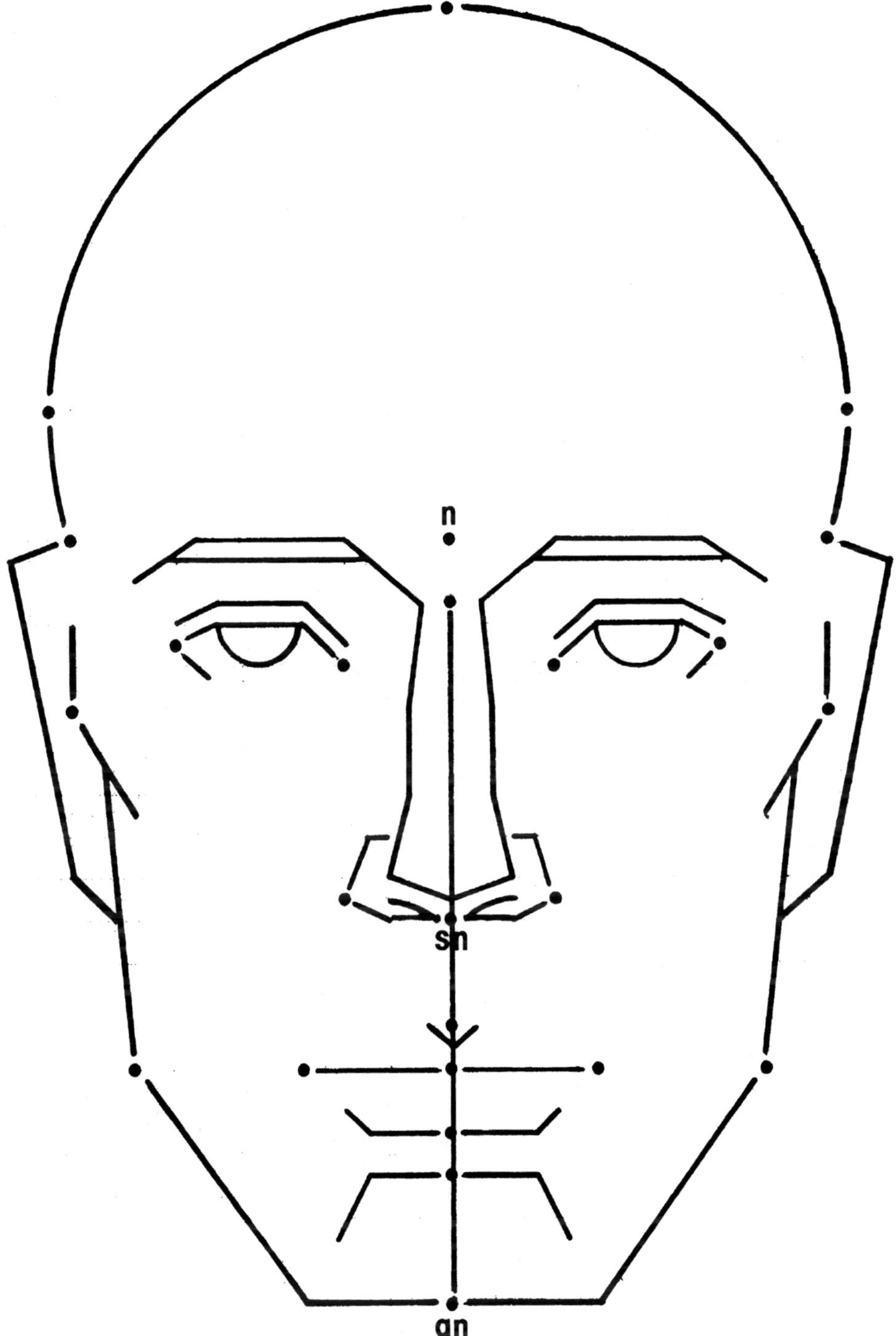

Figure 3.4. Nasal Length Index: n = nasion; sn = subnasale; gn = gnathion.

In some classical artistic canons of facial proportion, nasal length is one-half the facial length giving an index of 50.0 which is far longer than the average.

NASAL WIDTH INDEX

(al-al/n-sn × 100)

(Figure 3.5)

This index measures the width of the nose relative to its length.

	Narrow Nose	Normal Range	Wide Nose
Males	← 58.90	59.00–72.60	72.70 →
Females	← 59.20	59.30–69.40	69.50 →

These are ranges for the Caucasian nose. Ranges have not been established for peoples of Asian and African ancestry though they are known to be widest in the latter. In the author's unpublished data, the mean nasal width index for black males was 91.9 with a range of 76.0–105.0.

LABIAL INDEX

(ls-li/ch-ch × 100)

(Figure 3.6)

There appears to be very little data on this index which measures labial height relative to width. The ranges given here were taken from Olivier (1969) and were primarily used to define the major ethnic groups. Olivier found practically no sexual differences between European men and women.

Europeans (thin-lipped)—up to 34.9
Asians (medium thickness)—35.0 to 44.9
Africans (thick-lipped)—above 45.0

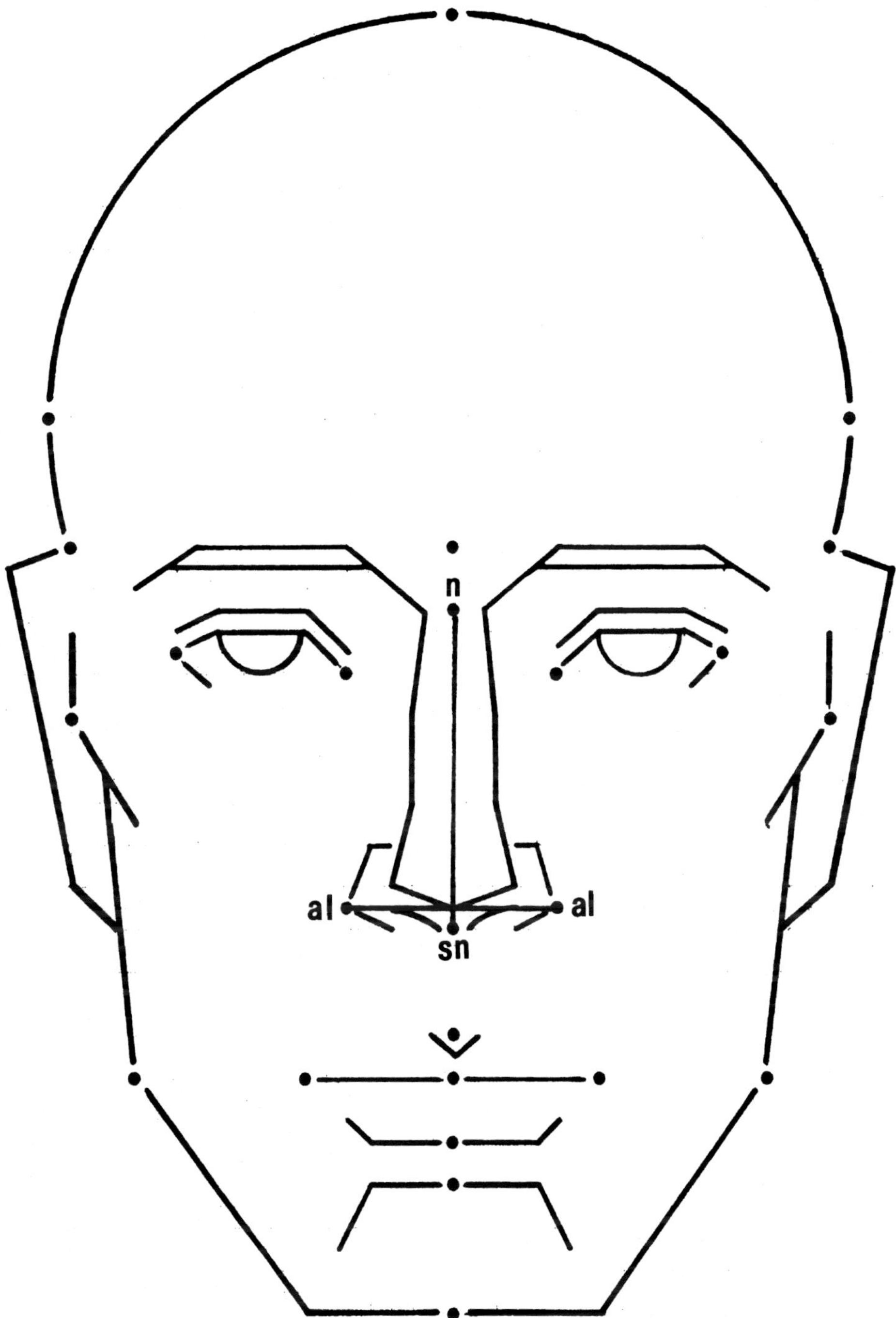

Figure 3.5. Nasal Width Index: n = nasion; sn = subnasale; al = alare.

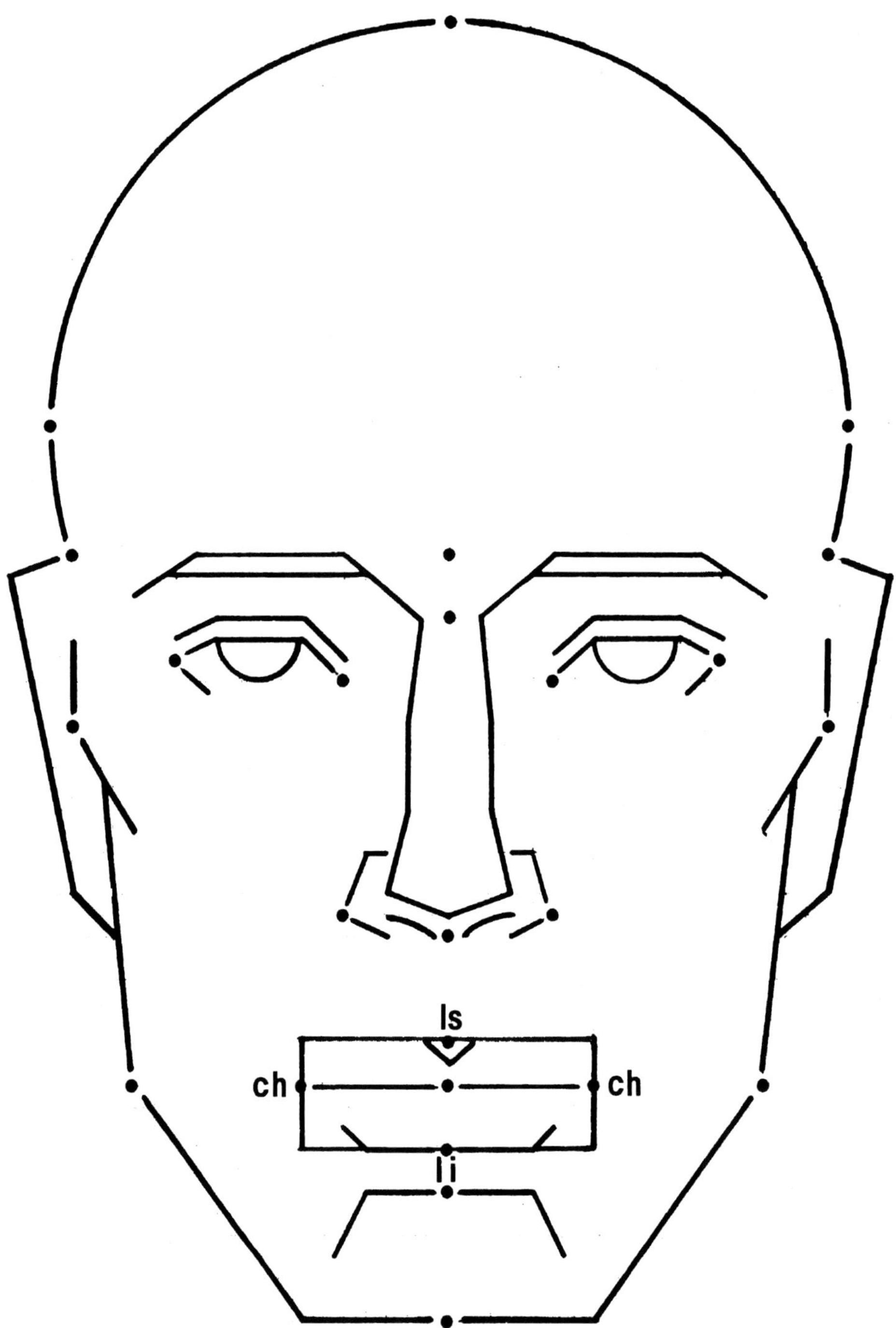

Figure 3.6. Labial Index: ls = labiale superius; li = labiale inferius; ch = chelion.

In actual practice, this is a very difficult index to assess due to the great range of variation within each of the above groups. In my unpublished Caucasian sample, lip eversion (exposure of the vermilion surfaces) ranged from almost non-existent (a straight labial fissure) to full ("voluptuous") with a high degree of overlap between the sexes.

UPPER LIP LENGTH INDEX

(sn-sto/sn-gn × 100)

(Figure 3.7)

This index measures the length of the upper lip relative to lower facial height.

These ranges were established by Farkas et al. (1984) based on a mean upper lip length of 32.4 percent in males and 31.1 percent in females.

	Short Lip	Normal Range	Long Lip
Males	← 28.00	28.10–35.50	35.60 →
Females	← 28.40	28.50–33.70	33.80 →

LOWER LIP LENGTH INDEX

(sto-lm/sn-gn × 100)

(Figure 3.8)

This index measures the length of the lower lip relative to lower facial height.

These ranges were established by Farkas et al. (1984) based on a mean lower lip length of 26.8 percent in males and 26.5 percent in females.

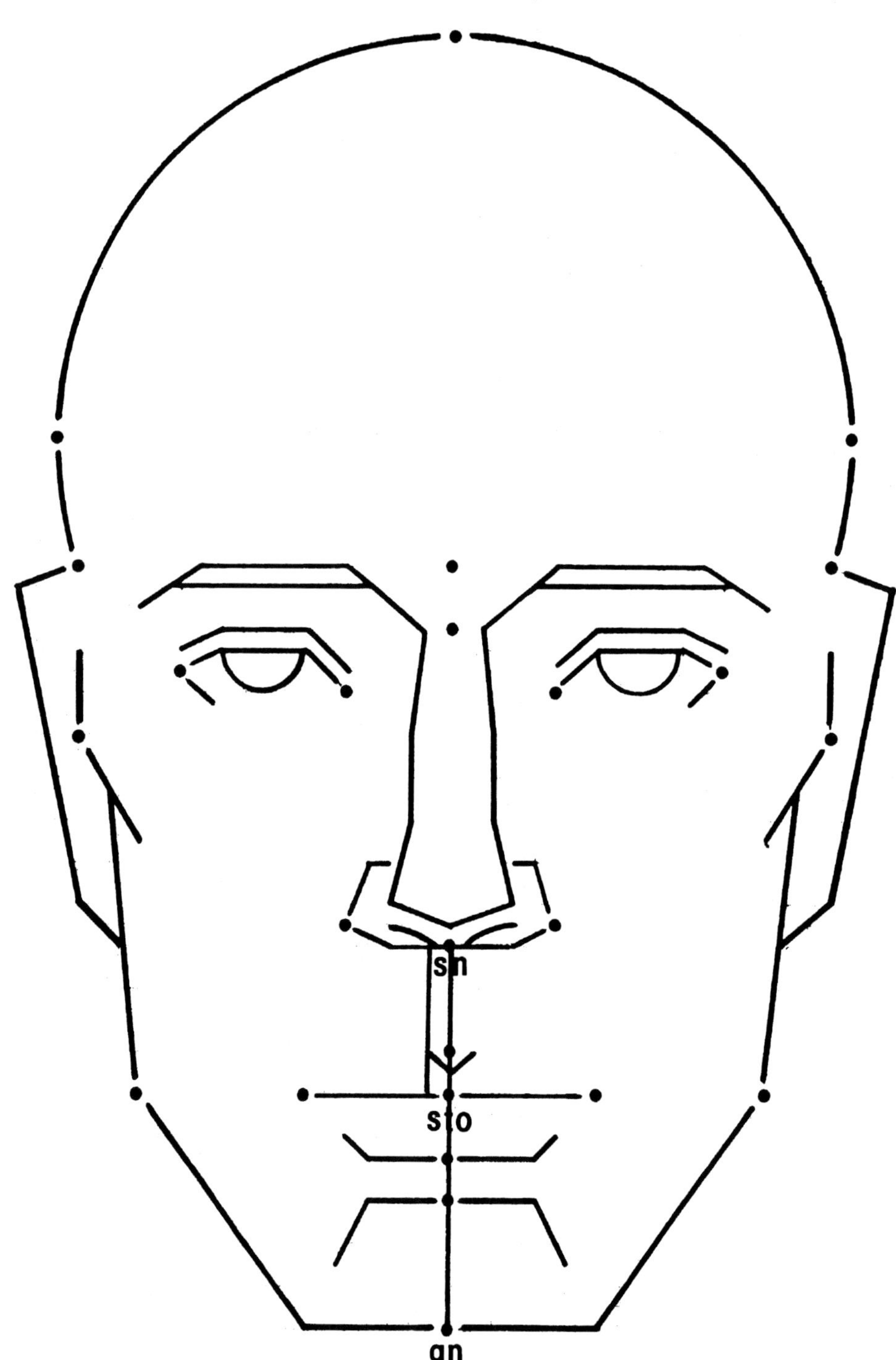

Figure 3.7. Upper Lip Length Index: sn = subnasale; sto = stomion; gn = gnathion.

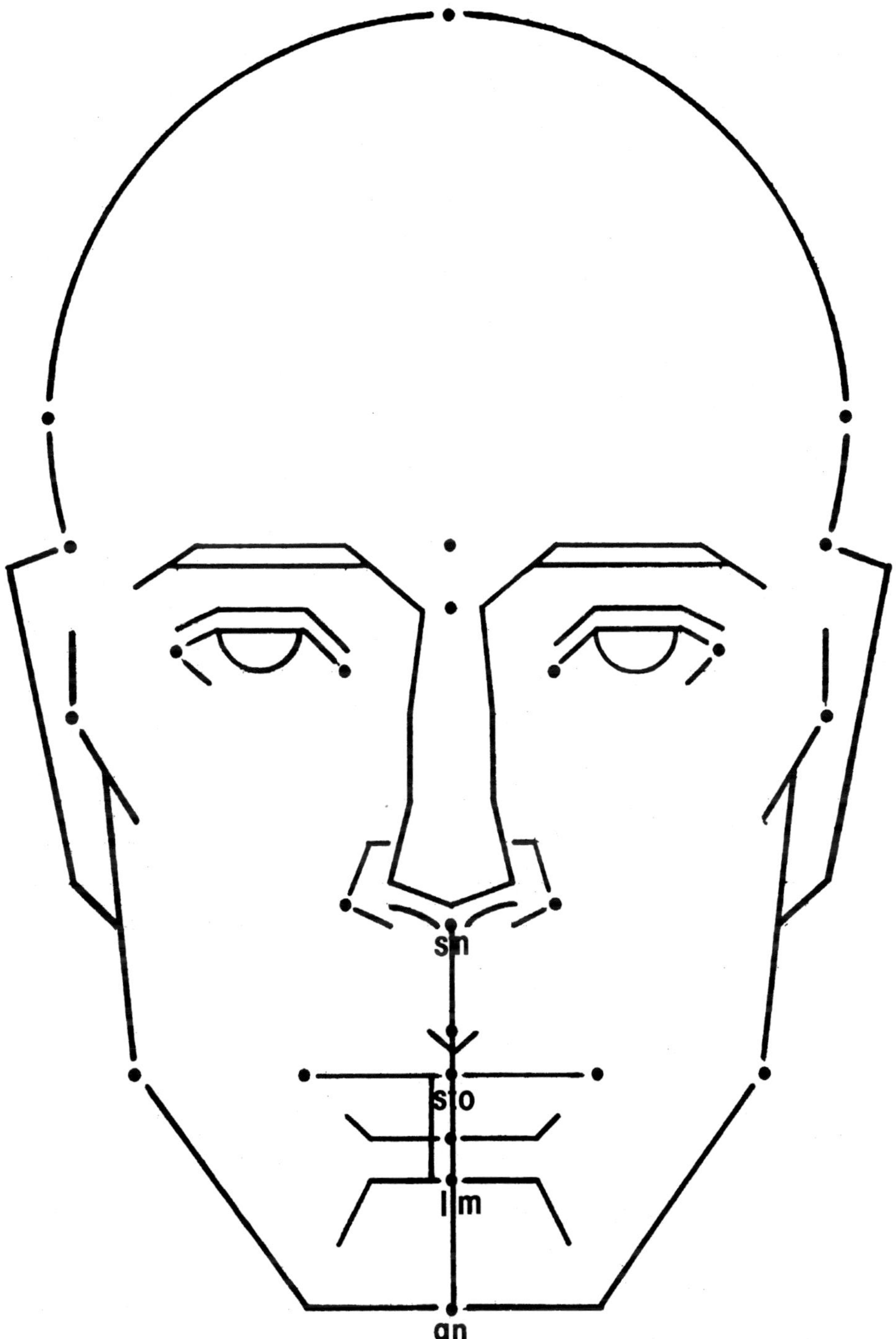

Figure 3.8. Lower Lip Length Index: sn = subnasale; sto = stomion; lm = labiomentale; gn = gnathion.

 Facial Geometry

	Short Lip	Normal Range	Long Lip
Males	← 23.70	23.80–29.80	29.90 →
Females	← 24.20	24.30–28.70	28.80 →

CHIN HEIGHT INDEX

(lm-gn/sn-gn × 100)

(Figure 3.9)

This index measures the height of the chin relative to lower facial height.

These ranges were established by Farkas et al. (1984) based on a mean chin height of 42.9 percent in males and 44.4 percent in females.

	Low Chin	Normal Range	High Chin
Males	← 38.10	38.20–47.60	47.70 →
Females	← 41.10	41.20–47.60	47.70 →

IRIDIO-CHELIAL INDEX

(im-im/ch-ch × 100)

(Figure 3.10)

This index measures the width between the irises relative to the width of the mouth and refutes the artistic canon that the width of the mouth is equal to the distance between the midpoints of the pupils. If this was true, the mouth would be excessively wide. I could not find a statistical evaluation for this index, but from my own unpublished observations a normal range of 93.3–107.7 seems reasonable (the lower the index, the wider the mouth). If im-im = ch-ch, the im-ch lines will be parallel. If the mouth is relatively wide, these lines will be divergent from im and if the mouth is relatively narrow, they will converge to-

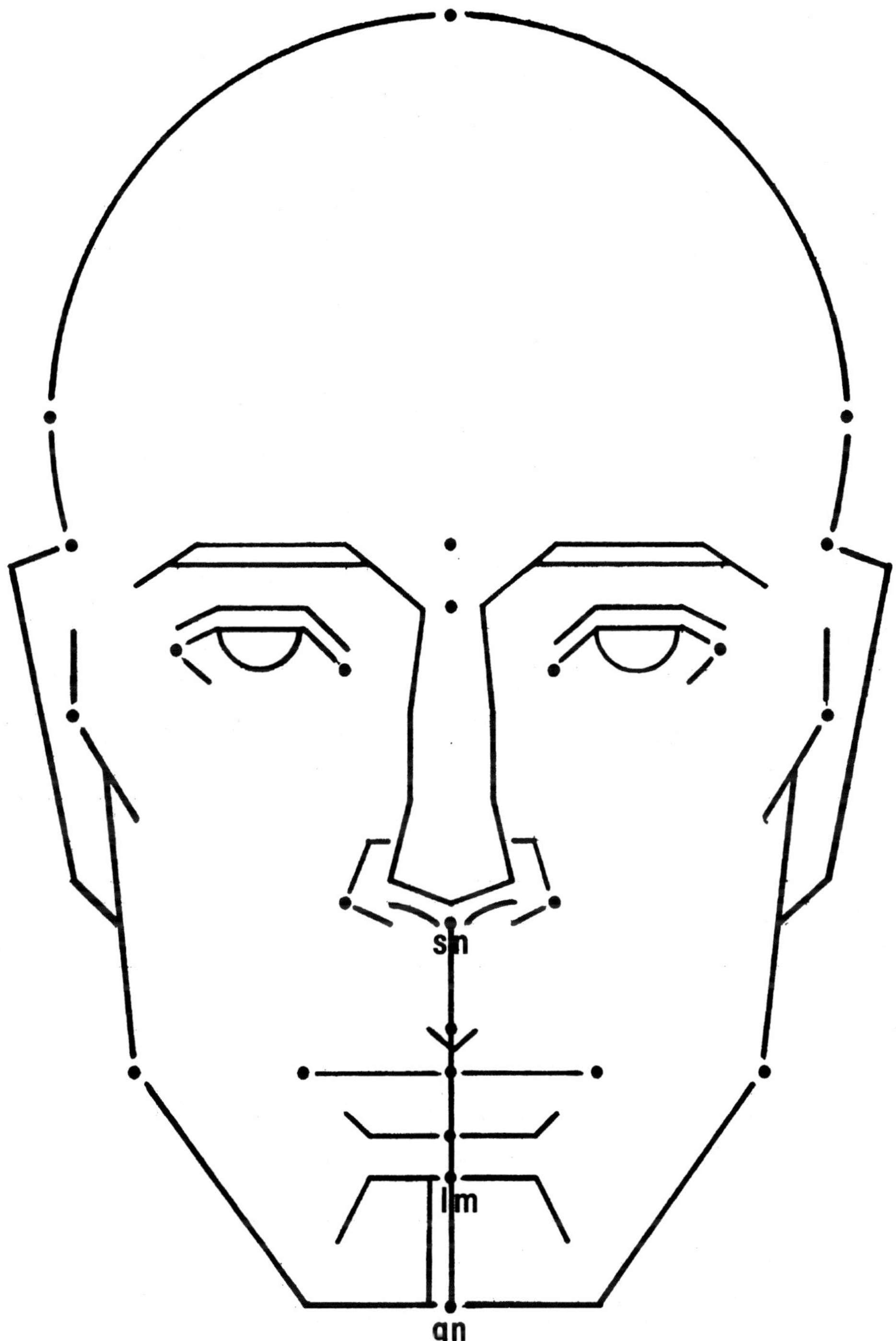

Figure 3.9. Chin Height Index: sn = subnasale; lm = labiomentale; gn = gnathion.

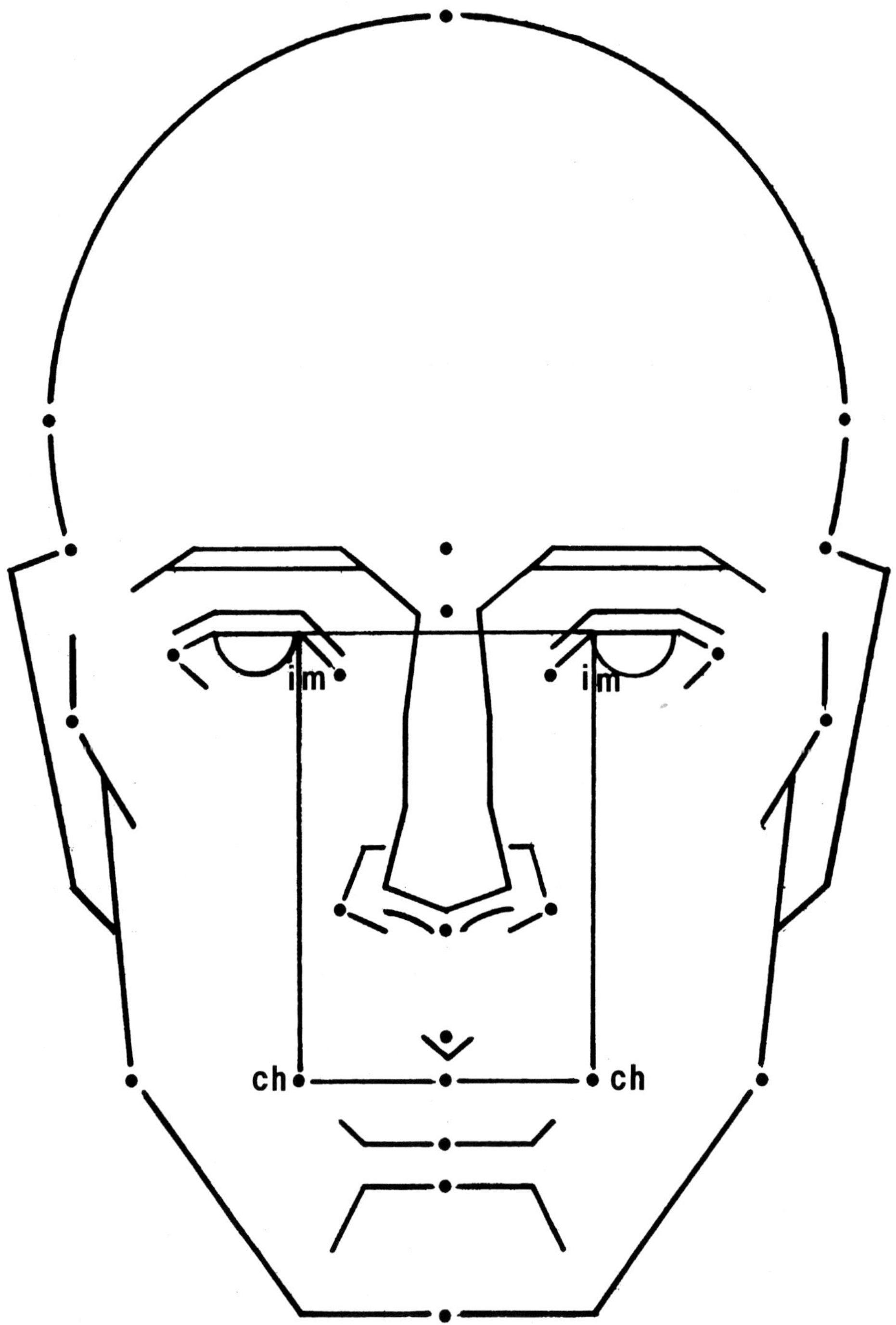

Figure 3.10. Iridio-Chelial Index: im = iridion mediale; ch = chelion.

ward ch. These lines appear to be more divergent in males and convergent in females (George, 1993).

ENDOCANTHAL-ALAR INDEX

(en-en/al-al × 100)

(Figure 3.11)

This index measures the width between the eyes relative to the width of the nose.

Farkas et al. (1985) found the intercanthal width to be equal to the nasal width in 40.8 percent of cases (n = 103). The nasal width was wider in 37.9 percent and narrower in 21.4 percent.

	Wide Nose	Normal Range	Narrow Nose
Males	← 87.10	87.20–103.00	103.10 →
Females	← 92.90	93.00–108.70	108.80 →

ALAR-CHELIAL INDEX

(al-al/ch-ch × 100)

(Figure 3.12)

This index relates the width of the nose to the width of the mouth.

Note that in "Phidias," the alar and chelial lines represent the parallel sides (bases) of an isosceles trapezoid. At 32°, the nasal triangle (ch-g-ch) is just slightly less than golden (see Figure 2.8).

	Narrow Nose	Normal Range	Wide Nose
Males	← 60.20	60.30–70.30	70.40 →
Females	← 58.10	58.20–68.30	68.40 →

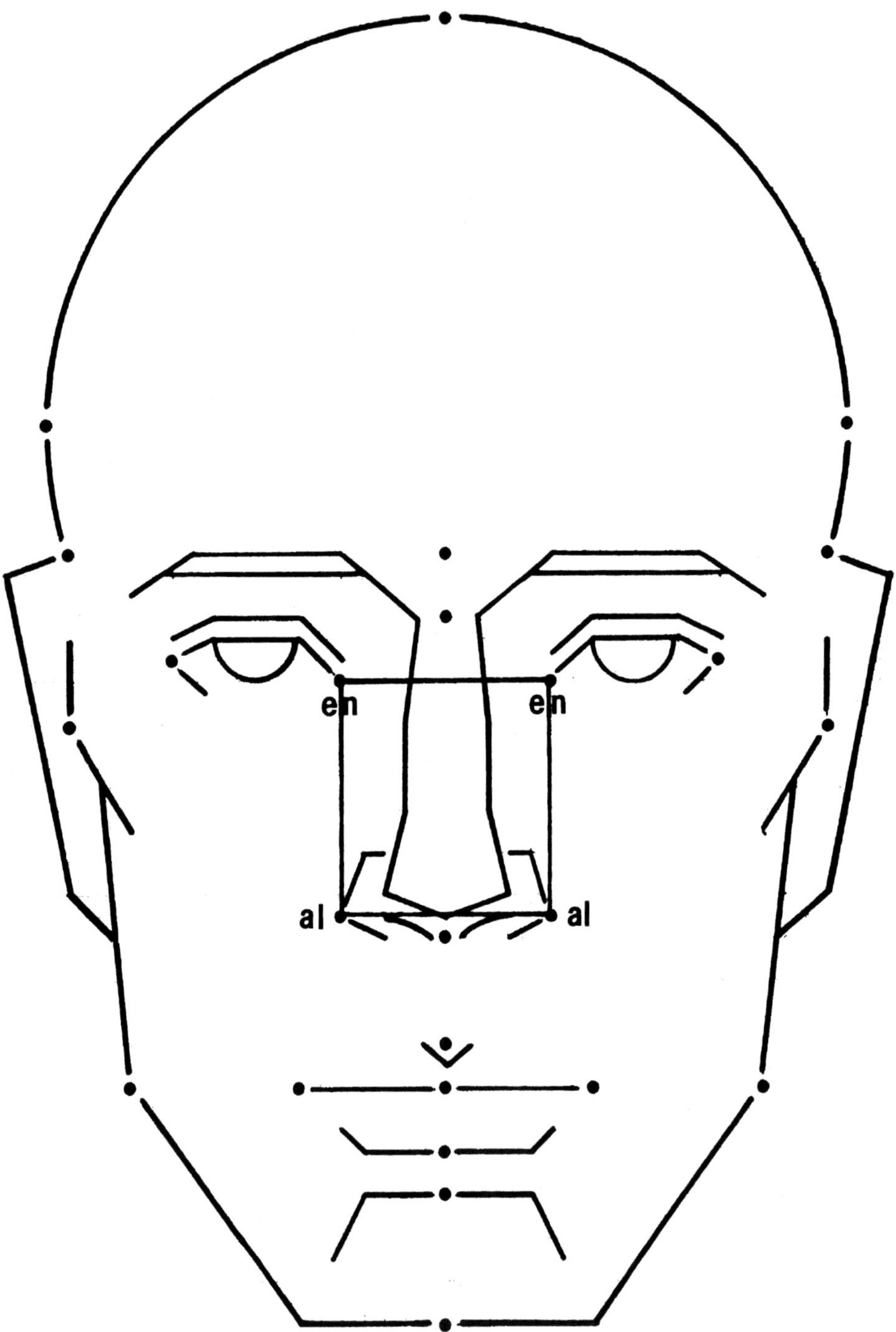

Figure 3.11. Endocanthal-Alar index: en = endocanthion; al = alare.

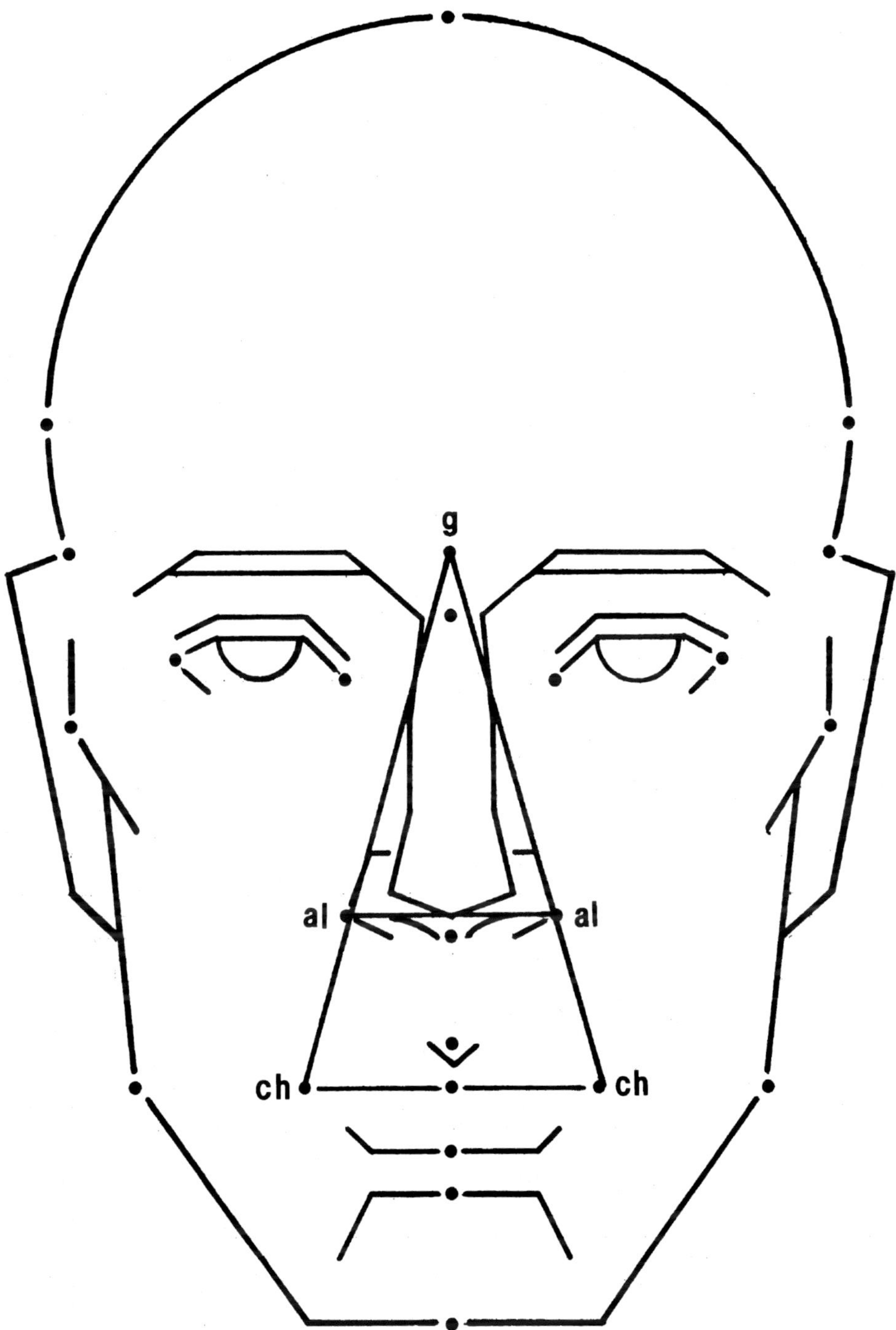

Figure 3.12. Alar-Chelial Index: g = glabella; al = alare; ch = chelion.

LABIO-ORBITAL TRIANGLE

(ec-li-ec)

(Figure 3.13)

This is the famous facial triangle that was illustrated in Figure 2.7. In graphic facial analysis it is another means of assessing eye width. It is distinguished by portrait artists as "The Facial Triangle" because in well-balanced faces it is a perfect equilateral triangle. This will, of course, convert to an index of 100 (ec-li/ec-ec × 100) since all sides will be equal.

NASO-ORBITAL TRIANGLE

(ec-sn-ec)

(Figure 3.14)

This is yet another measurement of ocular width. According to the properties of the facial square (Figure 2.5), this will be an isosceles right triangle. If the eyes are wide apart, the vertex (sn) angle will be greater than 90° and less than 90° if the eyes are close set. This is another useful relationship when trying to locate the lateral canthus in forensic facial approximations.

NASO-CHELIAL TRIANGLE

(ch-sn-ch)

(Figure 3.15)

This is another assessment of the width of the mouth and has the same properties as the naso-orbital triangle since both triangles form components of the "X" of the facial square. The wider the sn angle, the wider the mouth.

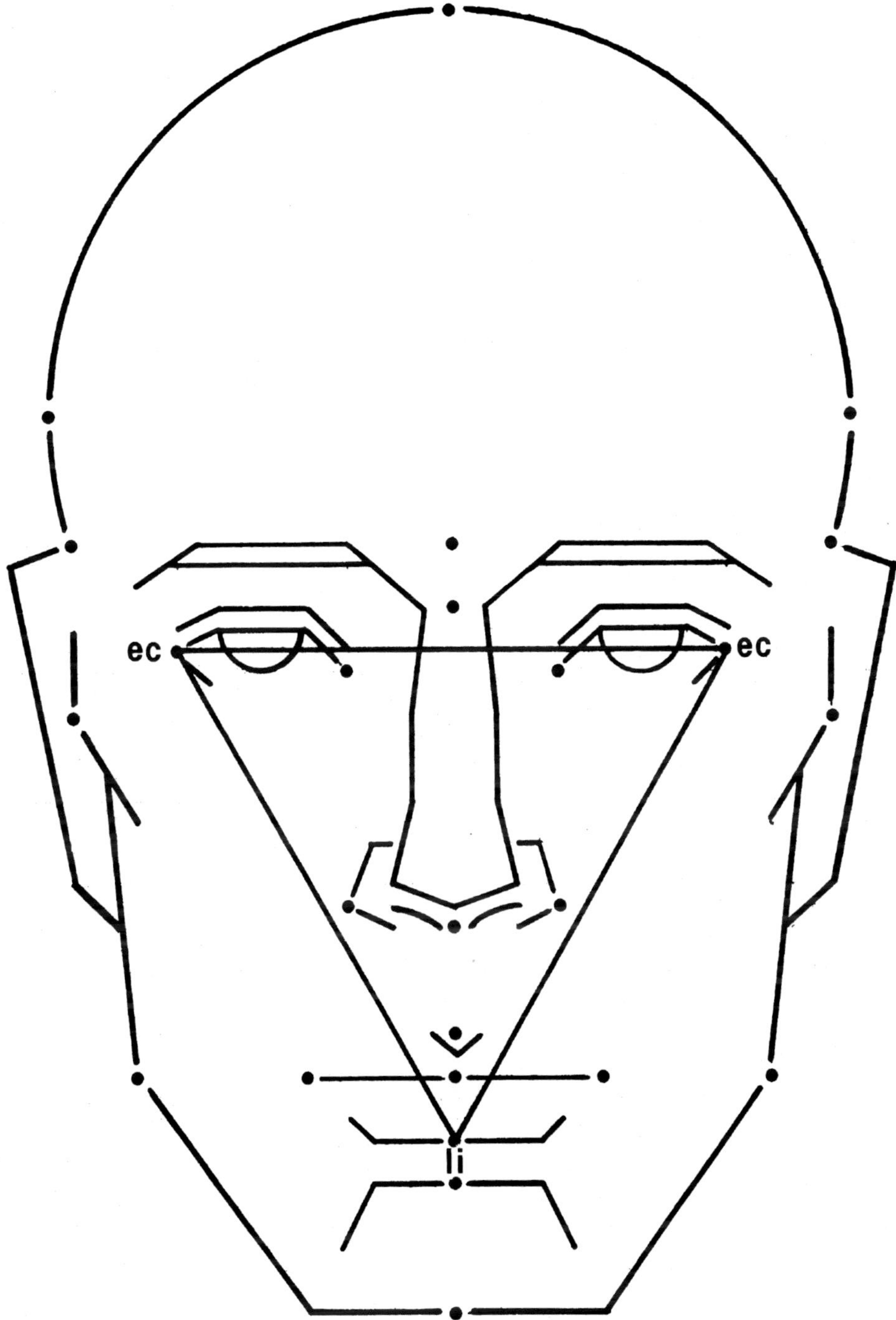

Figure 3.13. Labio-Orbital Triangle: ec = ectocanthion; li = labiale inferius.

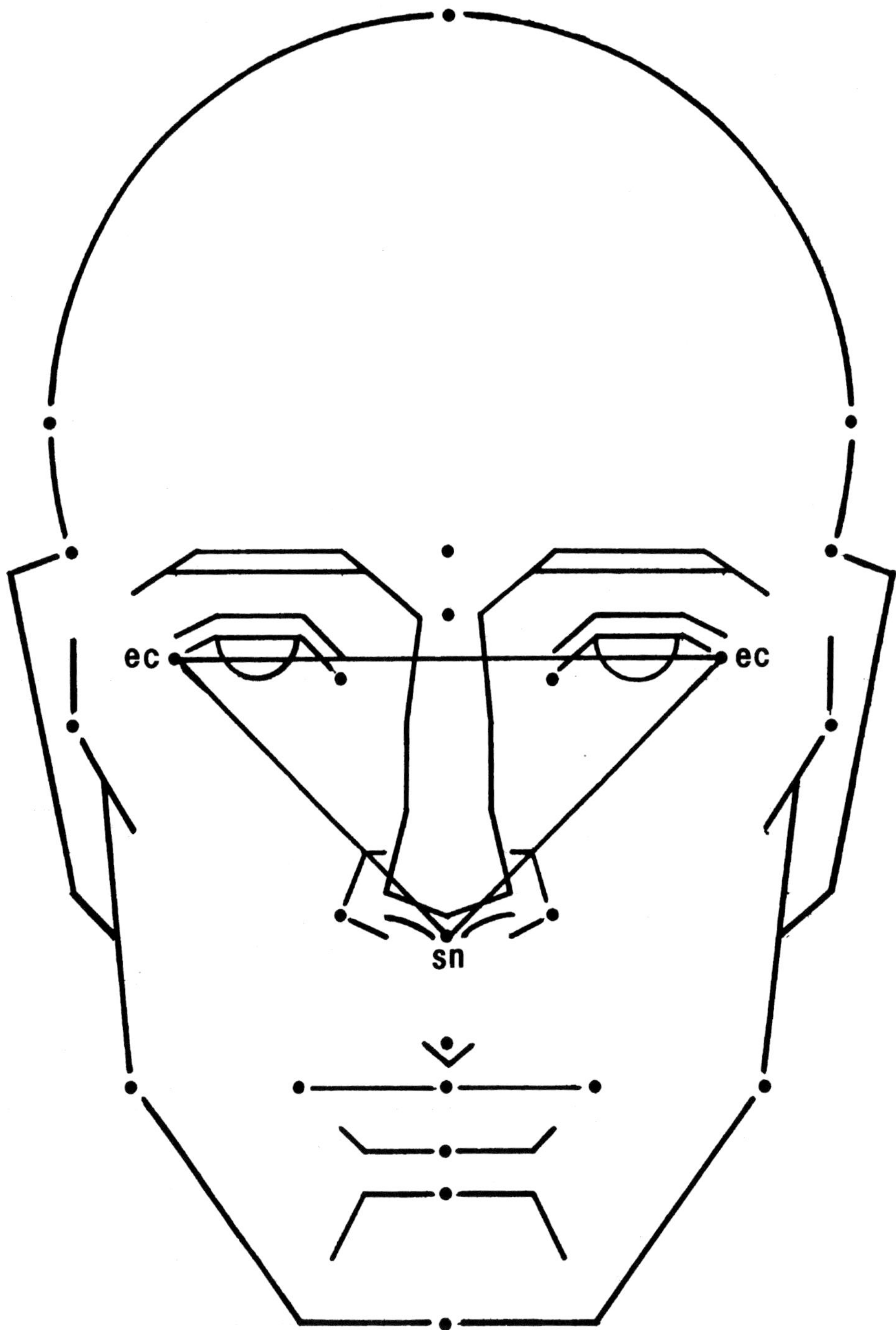

Figure 3.14. Naso-Orbital Triangle: ec = ectocanthion; sn = subnasale.

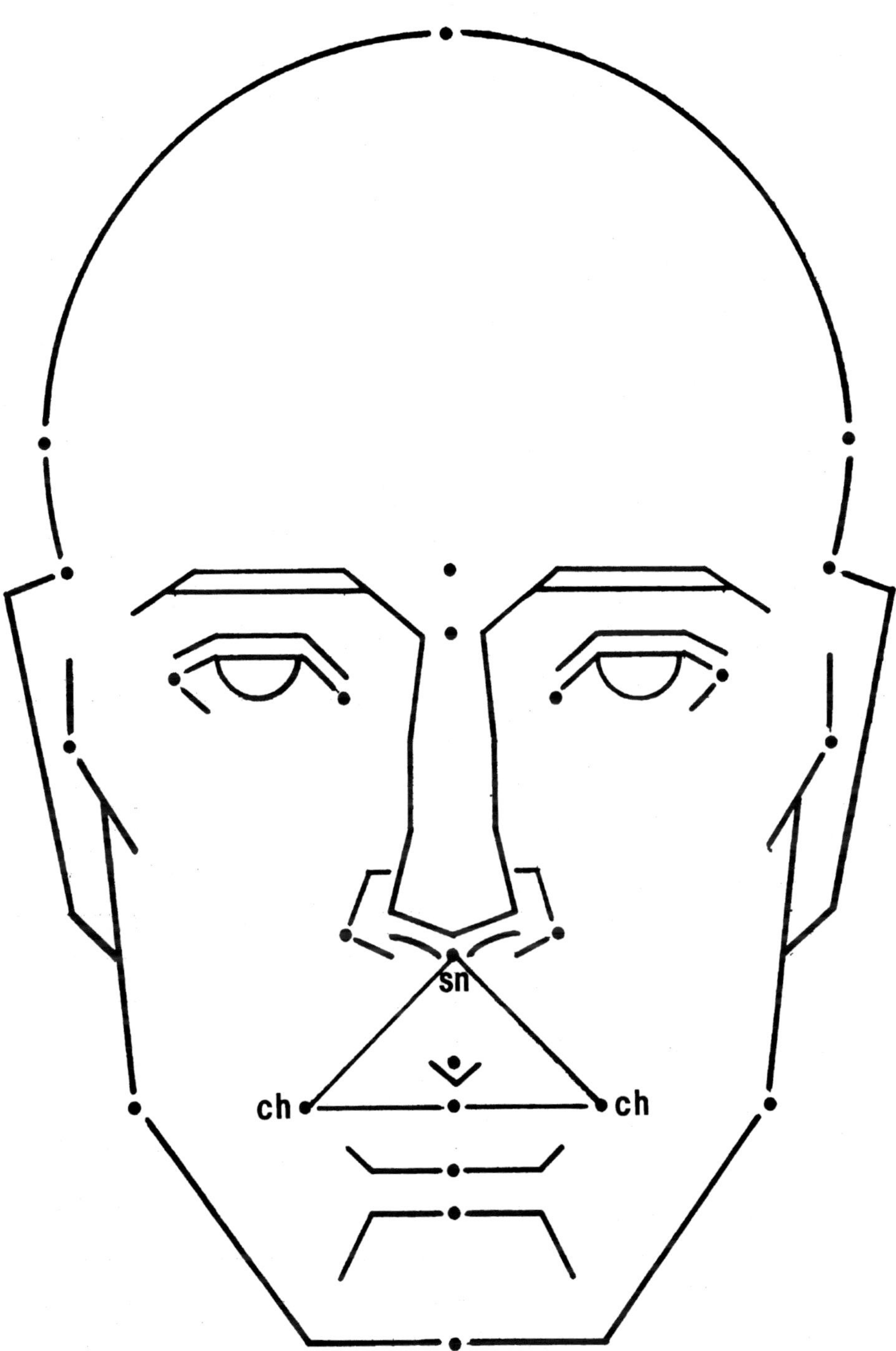

Figure 3.15. Naso-Chelial Triangle: sn = subnasale; ch = chelion.

FRONTAL ANALYSIS FORM

Table 3.1 is a frontal analysis form filled in for the unisex template used throughout this manual with data for Caucasian males. When substituting female means and ranges there are no significant differences. Phidias/Ophidia would therefore appear to be fairly average people with relatively wide jaws, long upper lips and low chins.

TABLE 3.1. FRONTAL ANALYSIS FORM

Case Number : __001__
Subject Description: (Sex) _?_ (Age) _?_
                       (Ethnicity) _Caucasian_

FRONTAL ANALYSIS

Index	<	Average	>
Facial Length		92.3	
Mandibulo-facial			83.5
Intercanthal		37.9	
Nasal Length		45.2	
Nasal Width		65.8	
Labial		35.7	
Upper Lip Length			39.6
Lower Lip Length		27.5	
Chin Height	33.0		
Endocanthal-Alar Index		100	
Iridio-Chelial		100	
Endocanthal-Alar		100	
Alar-Chelial			71.4
Labio-Orbital Triangle		60%	
Naso-Orbital Triangle		@ sn = 90°	
Naso-Chelial Triangle		@ sn = 90°	

Chapter 4

GRAPHIC FACIAL ANALYSIS:
LATERAL VIEW

PROFILE ANALYSIS: FACIAL ANGLES

The graphic analysis of the lateral face is largely a matter of angles relative to the following facial planes (Figure 4.1).

- *Frankfort Horizontal Plane* (FH)–This is a line connecting the tragion with the inferior orbital groove and represents the head in natural anatomical position. It is a basic plane of reference for many measurements in cephalometry and craniometry (= porion – orbitale).
- *Facial Plane* (FP)–This is a line connecting the nasion (n) and pogonion (pog).

Modern orthodontics owes a major debt of gratitude to the Dutch anatomist, Petrus Camper, who in 1786 published his description of the "facial angle" as a means of assessing degrees of facial prognathism (Finlay, 1980). Today, Camper's plane has been replaced by the Frankfort Horizontal and his angle has been superceded by the maxillary and mandibular angles.

It seems as if the number of angles employed by the modern orthodontist exceeds the number of stars in the Milky Way, but fortunately for the forensic artist only a few are necessary to evaluate a profile. The angles and indices selected for the following profile analysis are described in detail in *Proportions of the Aesthetic Face* (Powell and Humphreys, 1984) and *An Atlas and Manual of Cephalometric Radiography* (Rakosi, 1979).

 Facial Geometry

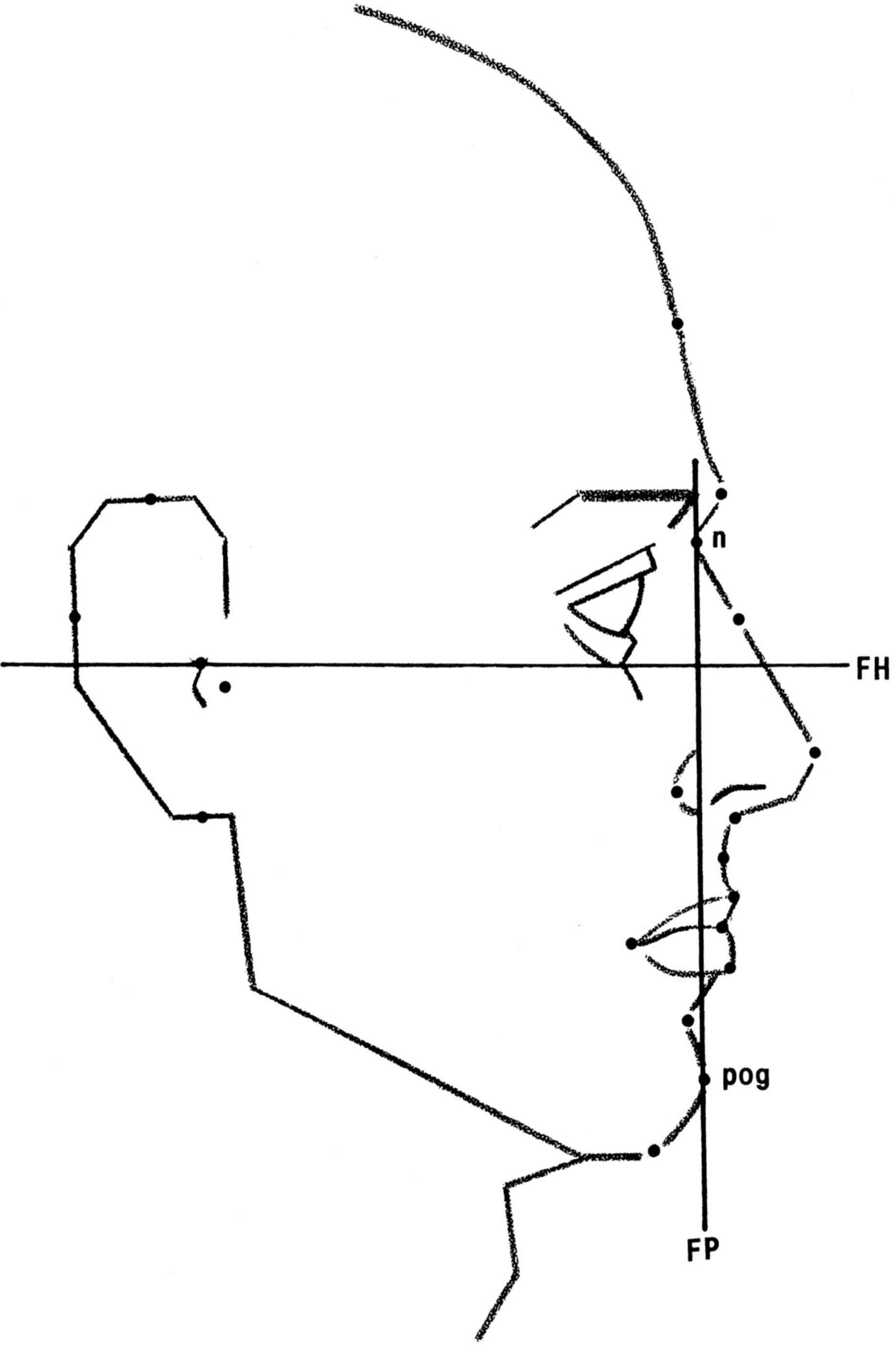

Figure 4.1. Facial Planes: FH = Frankfort Horizontal; FP = facial plane; n = nasion; pog = pogonion.

MAXILLO-FACIAL ANGLE

(n-sn @ FH)

(Figure 4.2)

The maxillo-facial angle is the posteroinferior angle of the intersection of line n-sn with the FH and measures the forward position of the maxilla. The greater the angle, the more protruding (prognathic) the maxilla. The lesser the angle, the straighter (more orthognathic) the maxilla. The esthetic range is 92°–100° with a mean valve of 96° (Holdaway, 1983).

Below 92°	92°–100°	Above 100°
retrognathic maxilla	orthognathic maxilla	prognathic maxilla

MANDIBULO-FACIAL ANGLE

(n-pog @ FH)

(Figure 4.2)

The mandibulo-facial angle is the posteroinferior angle at the intersection of the facial plane (n-pog) with the FH and measures the forward position of the mandible. The greater the angle, the more protruding (prognathic) the mandible. The lesser the angle, the more receding (retrognathic) the mandible. The range for a relatively straight (orthognathic) face is 84°–98° with a mean value of 91° (Holdaway, 1983).

Below 84°	84°–98°	Above 98°
retrognathic mandible	orthognathic mandible	prognathic mandible

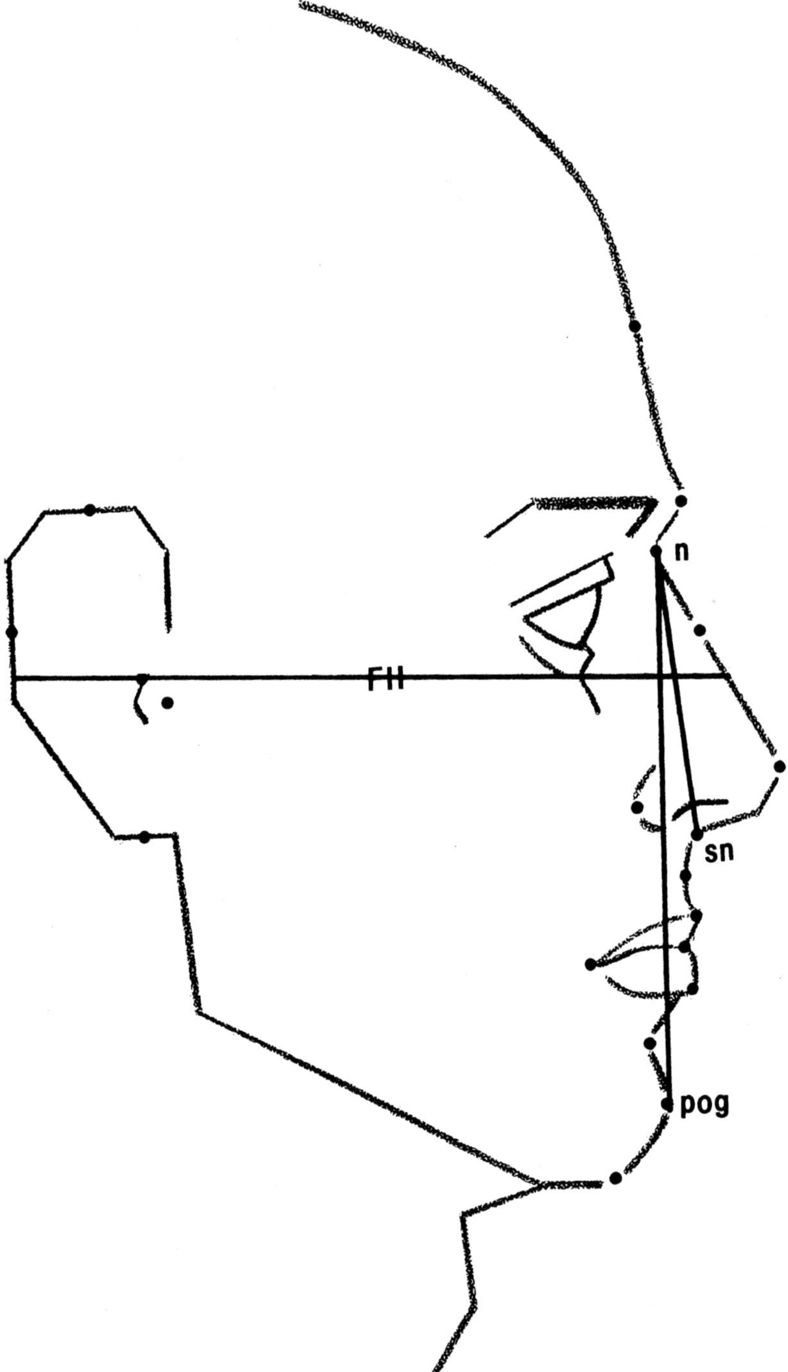

Figure 4.2. Facial Angles: FH = Frankfort Horizontal; n = nasion; sn = subnasale; pog = pogonion.

In an average orthognathic face the difference between the above two angles is approximately five degrees.

LATERAL FACIAL TRIANGLE

(n-prn-pog-n)

(Figure 4.3)

This triangle frames the nose and has the following sides and angles:

1. Facial plane (n-pog)
2. Nasal bridge line (n-prn)
3. Naso-mental line (prn-pog)–this is known as Ricketts' E (esthetic) line which is used in assessing lip relations (Ricketts, 1968).
4. Naso-facial angle (pog-n-prn)
5. Naso-mental angle (n-prn-pog)
6. Mento-facial angle (n-pog-prn)

This scalene triangle was named the "aesthetic triangle" by Powell and Humphreys (1984) who used it to evaluate patients and plan orthodontic and surgical treatments. They used the glabella point for the nasal plane instead of the nasion and considered the following ranges to be ideal:

1. Naso-facial angle   30°–40°
2. Naso-mental angle   120°–132°
3. Mento-facial angle   15°–25°

NASO-FACIAL ANGLE

(pog-n-prn)

(Figure 4.4)

As described in the previous section, the naso-facial angle (pog-n-prn) was incorporated into the "aesthetic triangle" by Powell and Humphreys (1984). These authors believed the ideal naso-facial angle to be 36 degrees with a range between 30 and 40 degrees. Lewis (1976)

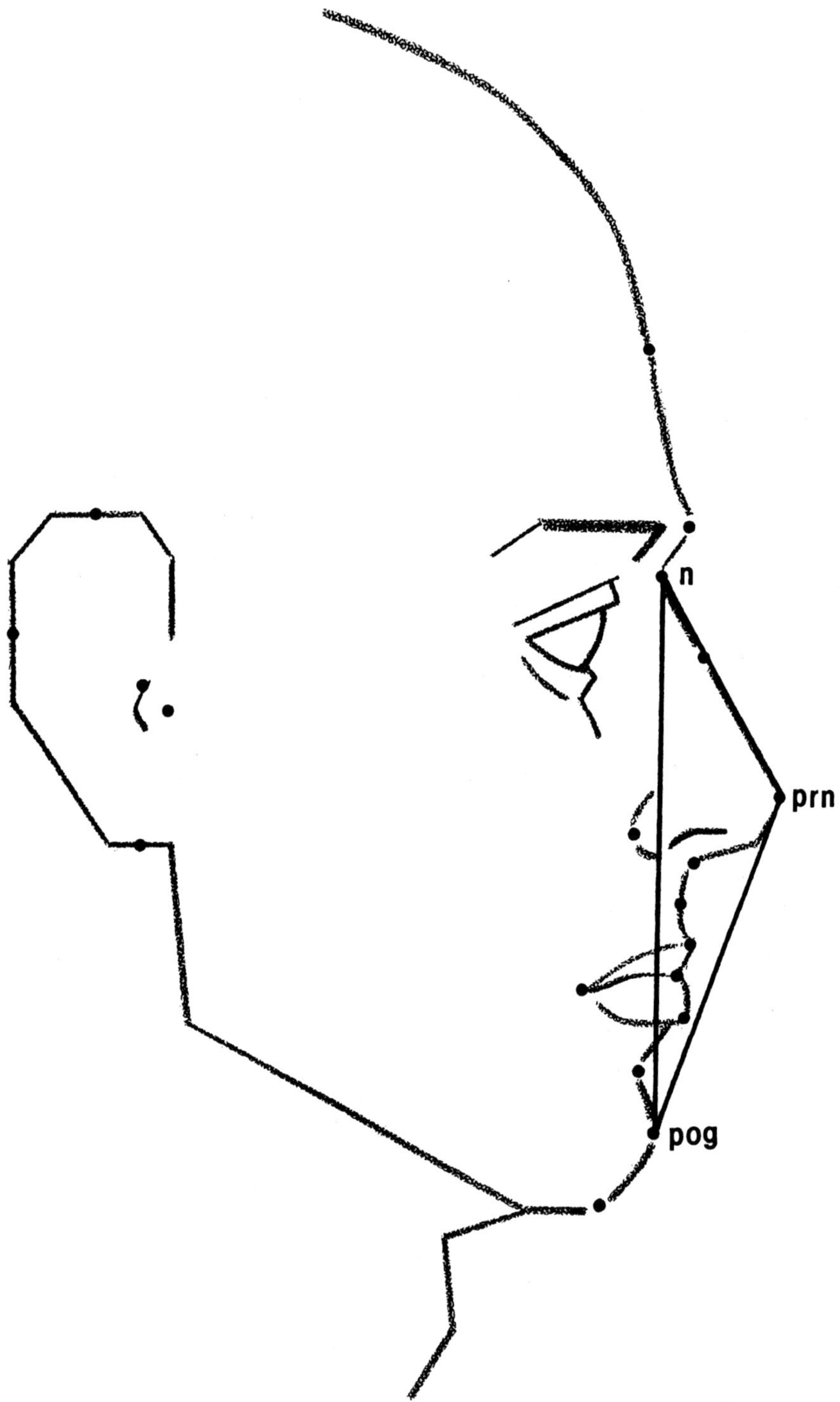

Figure 4.3. Lateral Facial Triangle: n = nasion; prn = pronasale; pog = pogonion.

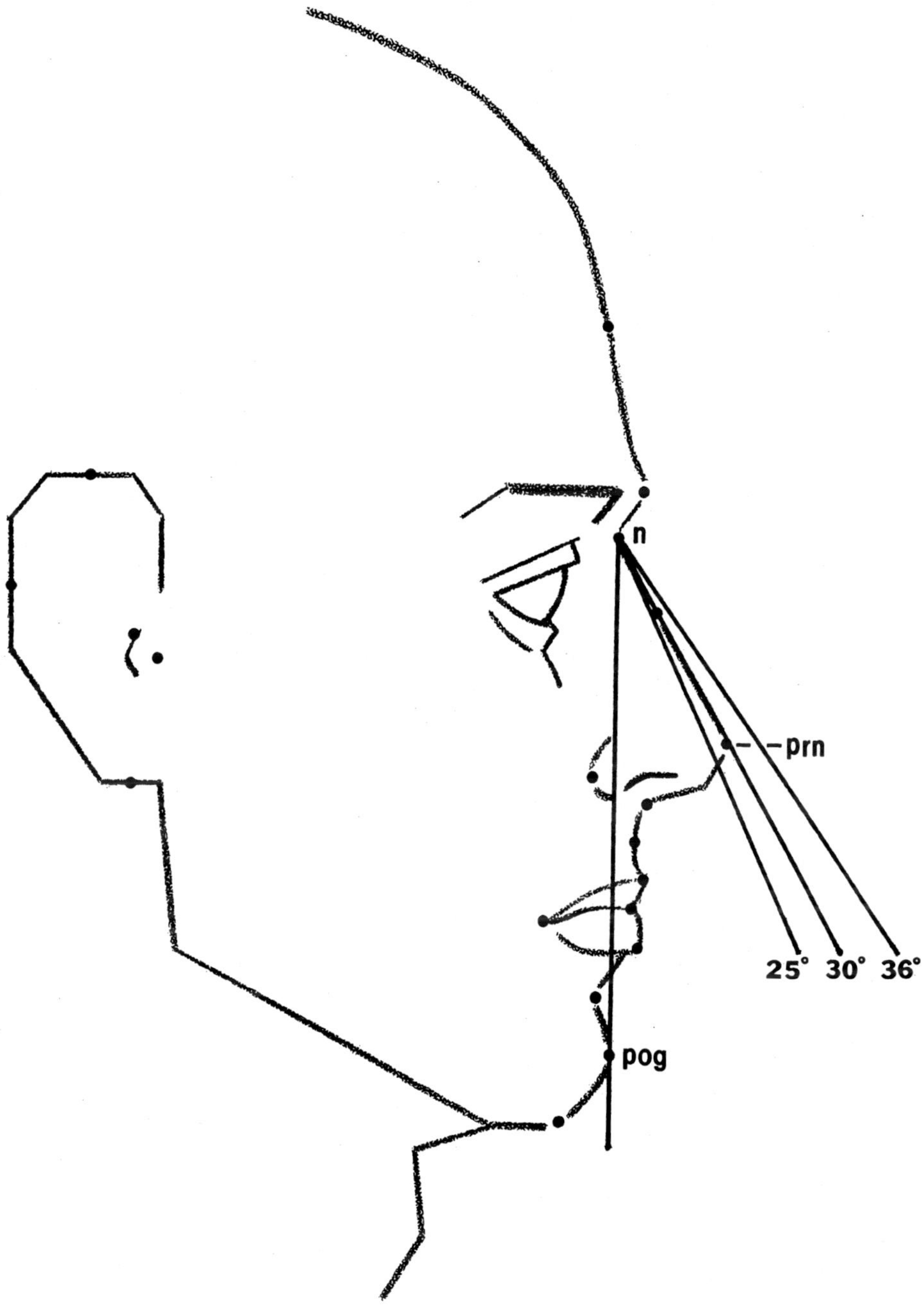

Figure 4.4. Naso-Facial Angle: n = nasion; prn = pronasale; pog = pogonion.

suggested a lower range of 25 to 35 degrees. The naso-facial angle is very useful in evaluating nasal projection, e.g., the shorter the nose, the lower the angle and vice versa. Thus, women tend to have lower naso-facial angles than men since statistically women have slightly shorter noses.

NASAL PROJECTION

(Figure 4.5)

Several methods have been devised for estimating and evaluating nasal projection (Powell and Humphreys, 1984). Rhinoplasticians use a nasal profilometer (rhinometer) for accurate measurements of the naso-facial angle, nasal tip angle and both vertical and horizontal nasal lengths (Lewis, 1976).

The basic shape of the nose is primarily determined by its vertical and horizontal lengths. The horizontal length (or width) is the distance between the alare and pronasale (al-prn). The vertical length is the distance between the nasion and subnasale (n-sn). The nasal index is calculated by HNL/VNL × 100. In my limited study of 17 males (ages 14 to 36) and 37 females (ages 14 to 34), I found the following ranges and means (George, 1987):

	Mean Index	Range
males	60.5	55–68
females	56.0	46–64

Simply stated, the HNL is approximately 60 percent (a "golden" index!) of the length of the VNL in males and is closer to 55 percent in females. This corroborates the ratios cited by Goode in Powell and Humphreys (1984).

NASAL TIP ANGLE

(Figure 4.6)

The nasal tip angle is formed by the intersection of the columellar line (nasal base) with the transnasal plane (TNP). On patients, this angle is measured with a profilometer. If the columellar line is below the TNP, the value is negative as in aquiline noses, and positive if above. Values of –20° to +40 have been recorded but the normal range falls between plus 5° to 25°. According to Lewis (1976), the esthetic range for males is plus 5°–15° and plus 10°–25° for females (line A in Figure 4.6).

Below 5°	5°–25°	Above 25°
low tip angles	esthetic range (males: 5°–15°) (females: 10°–25°)	high tip angles

In graphic facial analysis of photographs, this angle is often difficult to measure since the columellar line is normally curved rather than angled. A more accurate assessment of this angle would replace the columellar line with sn-prn which is easy to pinpoint (line B in Figure 4.6).

AURICULAR ANALYSIS

Like fingerprints, no two ears look alike and auricular variations have been used in criminal identifications (Swindler, 2004). While "forensic otology" is in its natal stage, requiring far more anatomical studies and statistical analyses (Swift and Rutty, 2003), most ears can be sorted by the auricular indices, the auricular inclination angle and qualitative gross anatomical differences. This is especially true when only two ears are being contrasted for identification as in photographic comparisons.

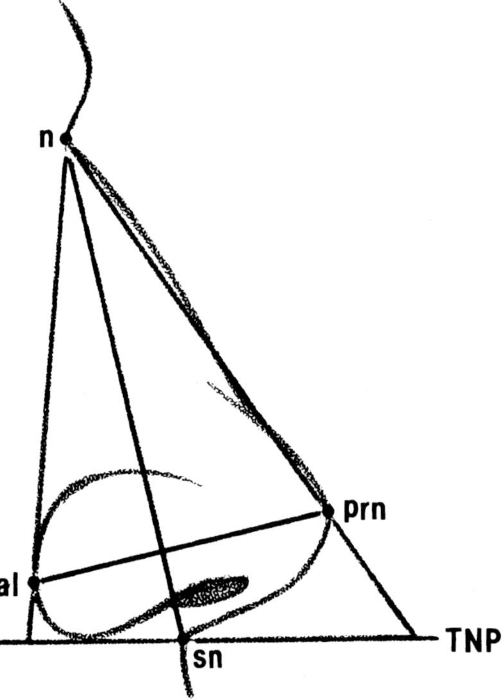

Figure 4.5. Nasal Projection: TNP = transnasal plane; n = nasion; sn = subnasale; al = alare; prn = pronasale.

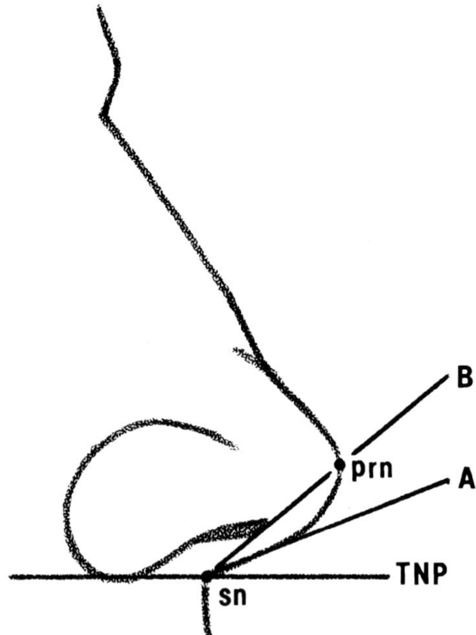

Figure 4.6. Nasal Tip Angle: TNP = transnasal plane; sn = subnasale; prn = pronasale.

EAR POSITION

(Figure 4.7)

In order to compare ear photographs, three important conditions must be met:

1. The ear must be photographed in lateral (profile) view.
2. The photos must be large and clear enough to be accurately measured.
3. It must be possible to establish the FH plane (po-or).

In most cases, these conditions are seldom met. The cephalometric points and external features of the ear necessary for the following measurements were defined and illustrated in Chapter 1 (Figures 1.2 and 1.9).

AURICULAR INDEX

(Figure 4.8)

The auricular index is derived by dividing the width of the ear by the length as follows:

1. Draw the FH (tragion-inferior orbital groove) and extend it posteriorly through the ear.
2. Draw a line parallel to the FH that is tangent to the superior rim of the helix. The point of contact is the superaurale (sa).
3. Draw a line parallel to the FH that is tangent to the earlobe. The point of contact is the subaurale (sba).
4. Draw line sa-sba. This is the axis of the ear and represents the length of the ear.
5. Draw a line perpendicular to FH and passing through the base of the tragus where it meets the posterior cheek. This point is the preaurale (pra).
6. Draw a line perpendicular to FH that is tangent to the posterior rim of the helix. The point of contact is the postaurale (pa).
7. The ear is now "boxed" in a rectangle with the width of the ear being the distance along the FH between the preauricular and postauricular lines.

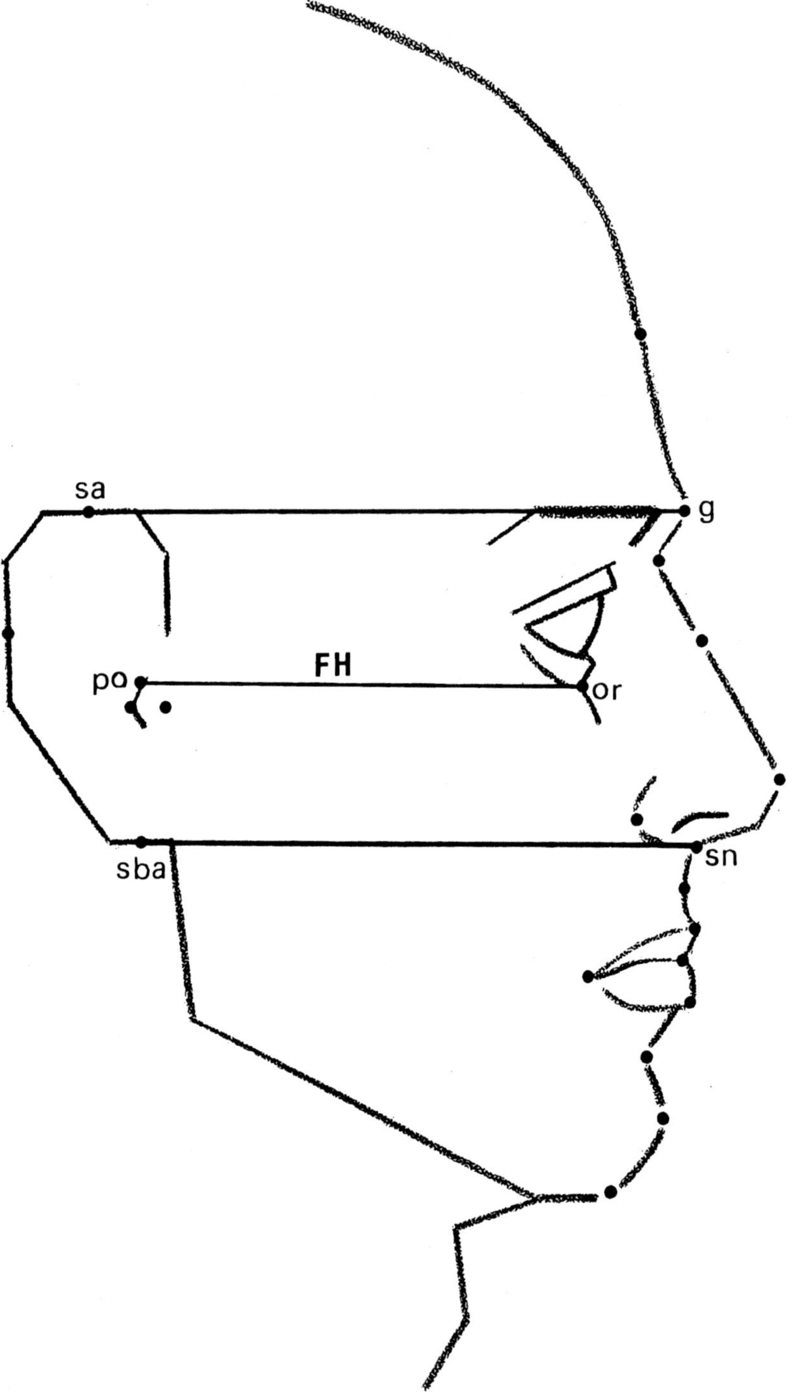

Figure 4.7. Ear Position: FH = Frankfort Horizontal; po = porion; or = orbitale; sa = super-aurale; g = glabella; sba = subaurale; sn = subnasale.

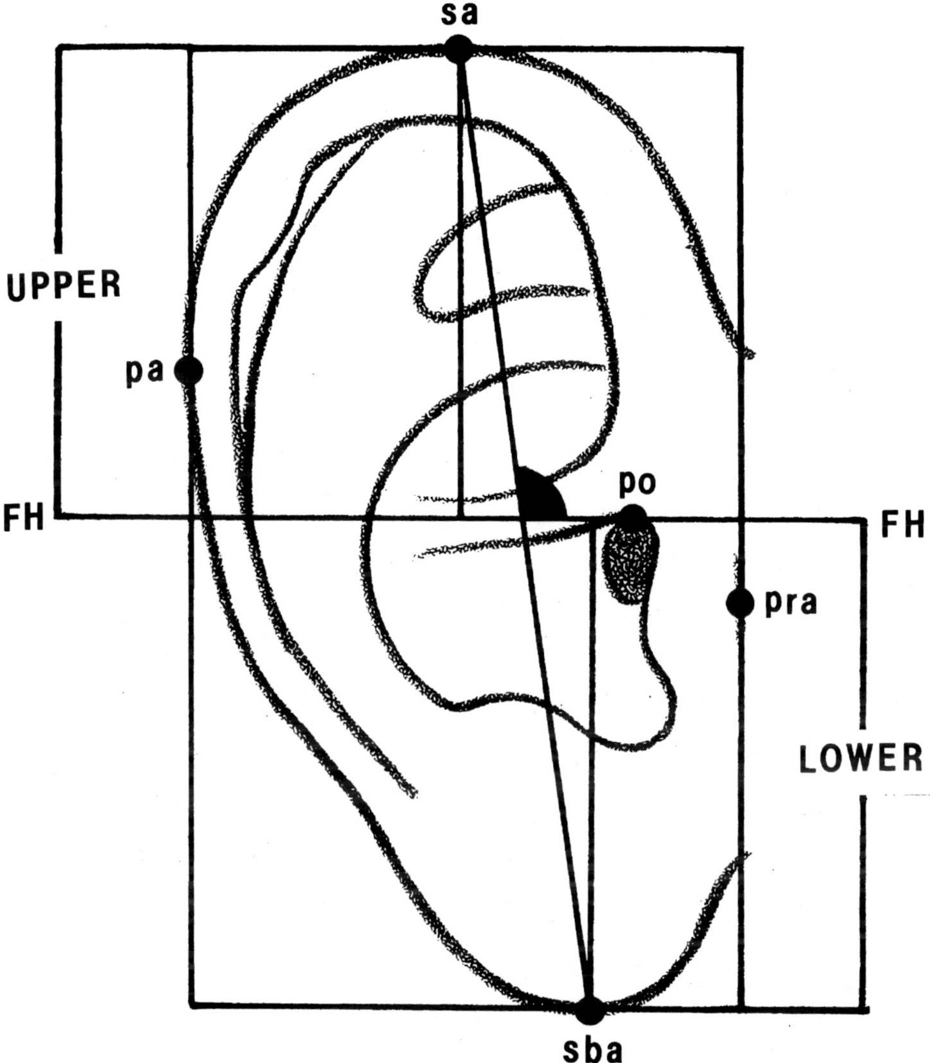

Figure 4.8. Auricular Index; Intra-Auricular Index; Auricular Inclination Angle; FH = Frankfort Horizontal; po = porion; sa = superaurale; sba = subaurale; pa = postaurale; pra = preaurale.

8. The auricular index may now be calculated by dividing the width by the length × 100. The following means and ranges were given by Farkas and Munro (1987):

	Mean Index	Range ±1 SD
Males	57.0	53.0 – 60.9
Females	57.4	53.6 – 61.2

INTRA-AURICULAR INDEX

(Figure 4.8)

This index determines whether the ear is "top heavy" or "bottom heavy." The upper half of the ear is measured along the axis from sa to its intersection with FH and the lower half is measured along the axis from sba to its intersection with FH. The top half is then divided by the lower half and multiplied by 100 to give the intra-auricular index. If the index is greater than 100, the ear is said to be top heavy and bottom heavy if less than 100. The ear in Figure 4.8 has an index of 96.2. This index has not been statistically evaluated but is nonetheless an excellent metric in photographic comparisons.

AURICULAR INCLINATION ANGLE

(Figure 4.8)

According to the neoclassical canons of facial proportion, the nasal bridge line should be parallel to the longitudinal axis of the ear. Farkas and Munro (1987) found this to be so in only 8.9 percent of their sample (n = 101). The longitudinal axis of the ear passes through sa and sba and the auricular inclination angle can be measured by the backward rotation of the axis at the FH. In Figure 4.8 the angle is indicated by the black quarter circle and measures 98°. Patterson and Powell (1977) considered the "esthetic mean" of the angle to be approximately 105°.

PROFILE ANALYSIS FORM

(Table 4.1)

Cephalometric angles and indices can be conveniently recorded on a profile analysis data form. This will give an overview of nasal and occlusal relations focusing attention on any paranormal features. In Table 4.1, a profile analysis has been completed for the drawing used to illustrate the profile relations discussed in this chapter. While aesthetically average for most relations, the nose tends toward the feminine side in projection and tip angle.

CONCLUDING REMARKS

Cephalometry has long been employed by orthodontists and facial surgeons for clinical assessments. Its relevance to forensic art, however, has been much less appreciated. Obvious applications would include photographic comparisons of suspects to mug shots or driver's license photos as well as photos taken at different ages, e.g., thirties versus fifties in cases of fugitives. Forensic artists may also be asked to compare authenticated photographs of historical figures with "newly discovered" photographs of the notable. I have done GFA's on photographs wishfully thought to be of Lincoln (the president) and Doc Holliday (the gunfighter) (George, 1989) with negative results in both cases. It must be emphasized that graphic facial analysis is basically an exclusionary technique. Similar GFA's are not proof that the subjects being compared are one and the same. However, if they are markedly different, the chances are so are the subjects in question (George, 2000).

ABOUT TEMPLATES

All forensic artists have a model face, blueprint or template in mind before beginning a composite sketch. This "template" is simply a rough plotting of the positions of the various facial features. The important point to remember is that this template is merely the starting point—not the finish line. It is meant to be modified as the witness' description unfolds. It is a cardinal sin to make the description conform to the template (Taylor, 2001).

TABLE 4.1. PROFILE ANALYSIS FORM

Case Number : __001__
Subject Description: (Sex) _?_ (Age) _?_
        (Ethnicity) __Caucasian__

Index	<	Average	>
Maxillo-facial Angle		98°	
Mandibulo-facial Angle		91°	
Naso-facial Angle	30°		
Nasal Projection	52.7°		
Nasal Tip Angle		17.5°	
Auricular Index		58.7°	
Intra-auricular Index		96.2°	
Auricular Inclination Angle		98°	

The "golden template" constructed for this manual (Figure 4.9) is not intended to be the blueprint of the universal human face. It simply demonstrates the fascinating geometric polygons that interrelate and humanize Caucasian facial features. In the interview, the witness will be asked if the nose is longer or wider, if the eyes are close together or farther apart, if the lips are thicker or thinner, if the face is narrower or chin broader, and so on. The finished composite drawing will certainly deviate to a greater or lesser extent from the "golden face." As demonstrated in Chapter 3, even "Phidias" deviates from the sample population in several facial relations.

Having thus qualified, justified and defended the construction of the golden face, the naturally curious will ask "So then, what did 'Phidias' look like?" One possible variant is illustrated by the masculine version in Figure 4.10 and another by the feminized version ("Ophidia") in Figure 4.11. Both appear to be reasonably normal individuals having been given reasonably normal facial features within the confines of the defined golden facial points. It is readily acknowledged, however, that myriad variations of the eyes, nose and mouth can be drawn connecting the given cephalometric points thus delineating faces that range from the sublime to those less comely. As astutely noted by Louise Gordon (1985), "No standard rules are included for those average heads which don't exist."

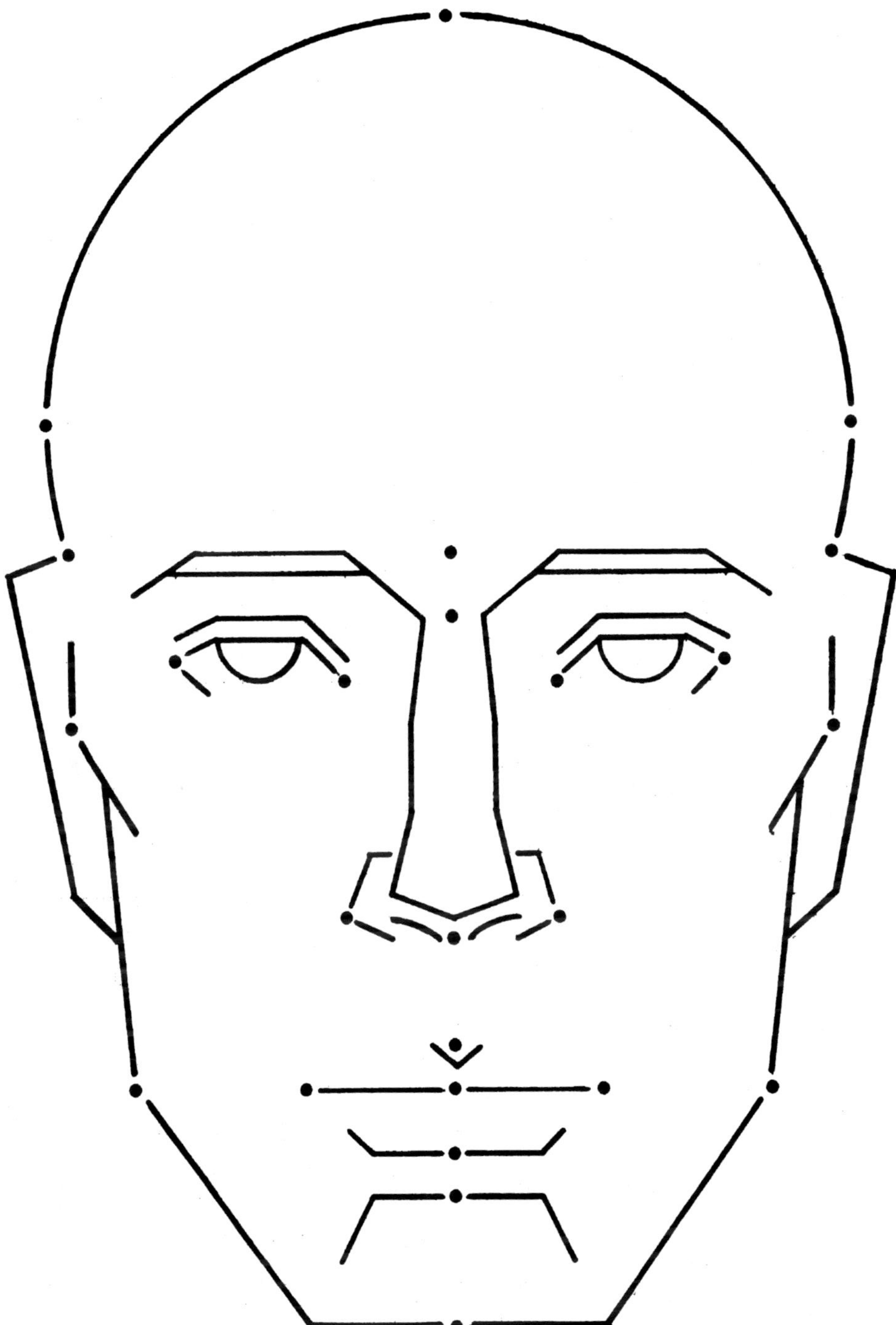

Figure 4.9. Golden Template.

 Facial Geometry

Figure 4.10. Phidias.

Figure 4.11. Ophidia.

LITERATURE CITED

Ashley-Montagu, M.F. (1935). The location of the nasion in the living. *American Journal of Physical Anthropology, 20* (1): 81–93.

Clement, John G., and Marks, Murray K. (2005). *Computer-Graphic Facial Reconstruction*. USA: Elsevier Academic Press.

Farkas, Leslie G., et al. (1984). Anthropometric proportions in the upper lip-lower lip-chin area of the lower face in young white adults. *American Journal of Orthodontics, 86* (1): 52–60.

Farkas, Leslie G., et al. (1985). Vertical and horizontal proportions of the face in young adult North American Caucasians: Revision of neoclassical canons. *Plastic and Reconstructive Surgery, 75* (3): 328–337.

Farkas, Leslie G., and Munro, Ian R. (1987). *Anthropometric facial proportions in medicine*. Springfield, IL: Charles C Thomas.

Finlay, Laetitia. (1980). Craniometry and cephalometry: A history prior to the advent of radiography. *Angle Orthodontist, 50* (4): 312–321.

George, Robert M. (1987). The lateral craniometric method of facial reconstruction. *Journal of Forensic Science, 32* (5): 1305.

George, Robert M. (1989). The Doc Holliday portraits: A simplified method of photographic comparison. Paper presented at the 41st Annual Meeting of the American Academy of Forensic Sciences, Las Vegas, Nevada.

George, Robert M. (1993). Anatomical and artistic guidelines for forensic facial reconstruction. In Iscan, Mehmet Y. and Helmer, Richard P. (Eds.), *Forensic analysis of the skull*. New York: Wiley-Liss.

George, Robert M., and Singer, Ronald. (1993). The lines and grooves of the face–A suggested nomenclature. *Plastic and Reconstructive Surgery, 92* (3): 540.

George, Robert M. (2000). Graphic facial analysis: The frontal view. Paper presented at the 9th Biennial Meeting of the International Association for Craniofacial Identification. Federal Bureau of Investigation, Washington, D.C., 2000.

Ghyka, Matila. (1946). *The geometry of art and life*. New York: Sheed and Ward.

Gordon, Louise. (1985). *How to draw the human head*. Great Britain: Penguin Books.

Hamm, Jack. (1982). *Drawing the head and figure*. New York: Perigee-Putnam.

Holdaway, Reed A. (1983). A soft tissue cephalometric analysis and its use in orthodontic treatment planning. *American Journal of Orthodontics, 84* (1): 1–28.

Huntley, H. D. (1970). *The divine proportion–A study in mathematical beauty*. New York: Dover.

Kayser, Alex. (1985). *Heads*. New York: Abbeville Press.

Kolar, John C., and Salter, Elizabeth M. (1977). *Craniofacial anthropometry.* Springfield, IL: Charles C Thomas.

Lewis, John R. (1976). The nasal profile (Chap. 18) in *Symposium on Corrective Rhinoplasty,* Vol. 13, D. Millard, ed. St. Louis: C. V. Mosby.

Livio, Mario. (2002). *The golden ratio.* New York: Broadway Books.

Marquardt, Stephen R. (2003). *Marquardt beauty analysis.* Multiple Web sites.

Olivier, G. (1969). *Practical anthropology.* Springfield, IL: Charles C Thomas.

Patterson, Carl N., and Powell, Donald G. (1977). Facial analysis in patient evaluation for physiologic and cosmetic surgery, in *Plastic and reconstructive surgery of the face and neck. Proceedings of the Second International Symposium, Vol. 1, Aesthetic Surgery,* G. A. Sisson and M. E. Tardy, Eds. New York: Grune and Stratton, pp. 146–159.

Powell, Nelson, and Humphreys, Brian. (1984). *Proportions of the aesthetic face.* New York: Thieme-Stratton Inc.

Rakosi, T. (1979). *An atlas and manual of cephalometric radiology.* London: Wolfe Medical Publications Ltd.

Ricketts, Robert M. (1968). Esthetics, environment and the law of lip relation. *American Journal of Orthodontics, 54* (4): 272–289.

Ricketts, Robert M. (1982). The biologic significance of the divine proportion and fibonacci series. *American Journal of Orthodontics, 81* (5): 351.

Shoemaker, William A. How to take the guesswork out of dental esthetics and function. *Florida Dental Journal,* Fall: 35.

Shoemaker, William A. (1987). How to take the guesswork out of dental esthetics and function–Part II. *Florida Dental Journal,* Winter: 25.

Swift, Benjamin, and Rutty, Guy N. (2003). The human ear: Its role in forensic practice. *Journal of Forensic Science, 48* (1): 153–160.

Swindler, Daris R. (2004). Personal Communication.

Taylor, Karen T. (2001). *Forensic art and illustration.* Boca Raton: CRC Press.

Wilkinson, Caroline. (2004). *Forensic facial reconstruction.* United Kingdom: Cambridge University Press, 2004.